The Indifferent Stars Above

*The Harrowing Saga of a
Donner Party Bride*

Daniel James Brown

An Imprint of HarperCollinsPublishers

HarperCollins books may be purchased for educational, business, or sales promotional use. For information please write: Special Markets Department, HarperCollins Publishers, 10 East 53rd Street, New York, NY 10022.

FIRST HARPERLUXE EDITION

HarperLuxe™ is a trademark of HarperCollins Publishers

Library of Congress Cataloging-in-Publication Data is available upon request.

ISBN: 978-0-06-177473-7

09 10 11 12 13 ID/RRD 10 9 8 7 6 5 4 3 2 1

For Sharon
Thank you

And they had nailed the boards above her face,
The peasants of that land,
Wondering to lay her in that solitude,
And raised above her mound
A cross they had made out of two bits of wood,
And planted cypress round;
And left her to the indifferent stars above.

—W. B. YEATS,
"A Dream of Death"

Contents

Author's Note

Even well after the tragedy was over, Sarah Graves's little sister Nancy often burst into tears for no apparent reason. She mystified many of her schoolmates in the new American settlement at the Pueblo de San José. One minute she would be fine, running, laughing, and playing on the dusty school ground like any other ten- or eleven-year-old, but then suddenly the next minute she would be sobbing. All of them knew that she had been part of what was then called the "lamentable Donner Party" while coming overland to California in 1846. Recent emigrants themselves, most of them knew, generally, what that meant and sympathized with her for it. But for a long while, none of them knew Nancy's particular, individual secret. That part was just too terrible to tell.

Nancy Graves's secret was just one part of many things that were too terrible to tell by the time the last survivors of the Donner Party staggered out of the Sierra Nevada Mountains in the spring of 1847. And for decades thereafter, many of those things were not told, except in tabloid newspaper accounts that were often compounded far more of fiction than of truth. It wasn't until a newspaper editor named Charles F. McGlashan began to delve into the story in the 1870s that many of the real details of what had happened that winter in the Sierra Nevada started to emerge. McGlashan set about interviewing survivors, and in 1879 he published his *History of the Donner Party*, the first serious attempt at documenting the disaster. Since then the true stories and the fictional ones have bred and interbred in the American imagination.

In introducing his recent book, *Patriot Battles: How the War of Independence Was Fought*, Michael Stephenson points out that our ideas about the generation that fought the American Revolution have become "embalmed" by "the slow accretion of national mythology." How true. I still cannot think of George Washington without visualizing him as a marble bust. But if the generation of 1776 has been fossilized by mythology, the same is equally true of their grandchildren. The emigrant generation of the 1840s has been endlessly depicted in film and television productions, almost always in highly stereotyped ways—a string of clichés about strong-jawed men circling the wagons to hold off Indian attacks and hard-edged women endlessly churning butter and peering out from under sunbonnets with eyes as cold and hard as river-worn stones.

The emigrants of the 1840s deserve better. They were, on the whole, a remarkable people living in remarkable times. Just how remarkable they were has largely been camouflaged for us, not only by the stereotyping and the mythologizing but by the homespun ordinariness of their clothes, the commonplace nature of their language, the simple virtues they held dear, and the casual courage with which they confronted long odds and bitterly harsh realities. The men among them did, in fact, sometimes circle their wagons to defend against potential Indian attacks, though the attacks were anticipated far more often than they ever occurred. But the men also lay awake at night agonizing about what they had gotten their families into, schemed to take advantage of one another, sobbed under the stars when their children died, lusted after women half their ages. And the women did, of course, churn a great deal of butter, but they also studied botany, counseled troubled teenagers, yearned for love, struggled with domineering husbands, and made wild rollicking love deep in the recesses of their covered wagons. Like all people in all times, the emigrant men and women, as well as the Native American men and women, of the 1840s were complex bundles of fear and hope, greed and generosity, nobility and savagery.

And in the end, each of them was, of course, an individual, as unique and vital and finely nuanced as you or me. So before I began to write this book, I made a vow to myself that wherever possible I would cut through the clichés and resist the easy assumptions about the men and women who went west in 1846. And I decided that I would focus on one woman in particular, Sarah Graves, and tell the unvarnished truth about what happened to her in the high Sierra in the terrible winter of that year.

Sarah hasn't made it easy. She left little record of her own experiences, and while others who suffered through the ordeal with her that winter have left us with their own, sometimes quite detailed accounts, few of them have had much to say about her. But what accounts there are suggest that she was a friendly, sociable, and thoughtful person, well liked by many of her companions but perhaps not apt to call attention to herself. And so while I have everywhere tried to remain entirely factual in writing about her, I have at times extrapolated from what those who accompanied her reported or from the published findings of experts in particular fields to describe what she must have experienced. Knowing, for instance, that she spent a specific night sleeping on the snow, clad only in wet flannel as the mercury plummeted into the low twenties, allows us also to know with reasonable certainty what she must have experienced in terms of physical discomfort, potential hypothermia, and psychological distress. So I have gone where eyewitness accounts and expert testimony have led me in describing such experiences. Similarly, in places I have drawn on my own experiences walking in her footsteps to re-create direct physical sensations that she must inevitably have felt. So, for example, in describing her passage through tall prairie grass, I have included sensory details from my own trek through the deep grass of the Willa Cather Memorial Prairie in Nebraska.

With that in mind, I offer this book not as a comprehensive history of the Donner Party but as a lens through which I hope you will be able to gaze with compassion and understanding on one young woman and all that the world once was to her.

DANIEL JAMES BROWN
Redmond, Washington
September 1, 2008

PROLOGUE

In many ways this book began one hot October afternoon in the fall of 2006 when I drove up the Napa Valley searching for bones. I have an affinity for bones. I like their hard-and-fast durability, the crispness of their lines, the heft and weight of them. Most of all I like their honesty. Bones have their secrets, but they tell no lies.

I made my way slowly up the valley in a rental car, trapped in the usual weekend procession of tourists promenading from one winery to the next on Highway 29. Heat waves rippled off of the black asphalt ahead of me. A vague autumnal haze hung over the valley, and so did the heady aroma of fermenting wine. The harvest was in full swing. Crews of pickers were making their way among the vines, stooped over, lugging white plastic boxes full of dark grapes. Watching them, I was glad for the car's air conditioner.

The wine tourists and I crept through the hot but picturesque old brick downtown of St. Helena, past the stately stone buildings of the Beringer winery, then past the Old Bale Mill. I craned my neck to see the mill, but its enormous wooden waterwheel was hidden behind a screen of redwoods. I remember the mill fondly as a shady, cool place my father used to bring me for picnics on warm days like this when I accompanied him on sales trips up the valley in the 1960s. Finally, just south of Calistoga, I pulled out of the parade and parked my car in front of a stately old farmhouse. The house is the headquarters for the Bothe-Napa Valley State Park now, but once it belonged to my great-uncle, George Washington Tucker.

I got out of the car. The house was closed for the afternoon, so I circled behind it and made my way up a little dirt path through a field of dried-out thistles and parched, waist-high grass toward what was left of a white picket fence. Sections of the fence had fallen to the ground, and the thistles had grown up tall and ragged between the pickets. Other sections still stood, leaning at odd, disjointed angles, flakes of white paint peeling from gray, weathered wood. From time to time, gusts of hot, dry wind blew down from the chaparral-cloaked Mayacamas Mountains just to the west, and the thistles rattled and scraped against the wooden pickets. Cicadas whirred in the big valley oaks fringing the field. The peppery scent of bay laurel spiced the wine-rich air.

The tumbled-down fence surrounded the remains of a derelict cemetery. Most of the tombstones were made of white marble, and many of them were cracked. Some had toppled to the ground and lay facedown among the weeds; others were missing essential parts, bearing only fragmentary inscriptions half obscured by patches of gray-green lichen. I began to wonder if I would find the one I was looking for. But then I did. Chiseled into the marble was the inscription GEORGE W. TUCKER DEC. 15, 1831–AUG. 16 1907.

He was my father's uncle, and I was there because he is my one tenuous connection with a young woman named Sarah Graves Fosdick. I had come to commune with his bones before I began to search for hers. He was not a major figure in Sarah's life, but once, a long time ago and far from there, he had been in her company. He was only fifteen then, she twenty-one, and they had both been setting forth on a journey of staggering proportions into, quite literally, uncharted territory. Whether they were even acquainted in the first few weeks of that trip, I do not know. They had both begun their journeys in Illinois—he in Illinois City and she in Sparland, near Peoria. A month later, on the west bank of the Missouri River, their families had met up and become friendly. It's unlikely that she would have paid him, a mere boy, any particular heed. But I do know that they must eventually have come to know each other, thrown together by circumstances so horrifying that neither of them could have imagined anything of the kind when they first met. He had traveled where she traveled, saw and felt much of what she saw and felt. His bones had walked the same tedious trails, stretched out at night on the same patches of prairie sod, climbed laboriously through the same icebound mountain passes.

I am far from the first to commune with bones in hopes of understanding Sarah's story. One warm August afternoon in the summer of 1849, Wakeman Bryarly, a twenty-eight-year-old doctor from Baltimore, found himself near what was then still called Truckee Lake in the high Sierra. Like so many other ambitious young men that summer—the summer when a whole world of young men seemed to pour across North America and into California—he was on his way to the goldfields. His party had encamped just east of the lake, and, with an afternoon to kill, he decided to take the opportunity to indulge in a cold bath. On his way to the lake, he hoped also to find something else that hundreds of other travelers that summer had sought out—a local tourist attraction of sorts.

He set off on foot, and just 150 yards down the road he found the first evidence of what he was looking for. In a dusty meadow full of whirring grasshoppers, dry grass, and foot-tall plants with broad gray leaves called "mule ears" stood a weathered but neatly fashioned log cabin. The cabin was surrounded by some unusually tall stumps, the remains of pine trees that had been cut off ten feet or more above the ground. He examined the cabin and found that it had two entrances and two living chambers separated by a log partition. In the dirt floor of each chamber, there was a shallow depression, the remnants, perhaps, of fire pits, or burial pits of some sort. Poking about in the dry grass among the stumps outside the cabins, he found some charred logs. And then, nearby, he found what he'd been told he would.

Half hidden in the grass were piles of bones. At first most of them seemed to be the bones of cattle, but then, just to the left of these, he found a nearly complete human skeleton sprawled out on the ground with grass growing up between the ribs. He stooped and examined the remains. Then he noticed that in the grass nearby there were bits and pieces of broken wooden boxes and some faded articles of clothing. He picked up a child's stocking and felt something rattling around inside it. He carefully turned the stocking inside out and dumped its contents into his hand—the small and perfect foot bones of a child.

Bryarly stood up and contemplated the scene before him. As a doctor, he had seen plenty of human bones before. And just recently he had seen a great many. The overland trail on which he'd been traveling all summer had been strewn with shallow graves, as many as ten per mile in some places. Most of them were the product of the devastating Asiatic cholera epidemic that had spread its way from New Orleans up the Missouri River and then westward along the trail that summer, killing hundreds of emigrants of all ages. The graves were often shallow, dug in sandy soil by grim-faced men anxious to move on and escape whatever mysterious agent was causing all these deaths. Many of the graves had been ripped open by wolves and coyotes, and in spots the road had been strewn with bones and tattered bits and pieces of mummified corpses.

Satisfied that he had seen what he'd come to see, Bryarly shook off the grimness of his discovery and continued along a pretty little creek through sparse pinewoods to Truckee Lake. There had been frost on the grass that morning, but the afternoon had grown warm and dry, as they often do on the east side of the Sierra in late summer. He was dusty and travel-weary, so he undressed and took a bath. The sheer beauty of the place stunned him. He exulted in the clarity of the lake's frigid, crystalline water and in the drama of the stark granite cliffs and basalt crags towering above the far end of the lake.

Exhilarated, he started back toward his party's encampment. And then, unexpectedly, as he passed through some dark pinewoods, he came across more evidence of what had happened there just three years before—the remains of another cabin, this one burned to the ground. Among blackened logs he found more human bones—many more. He crouched to examine them, and this time the bones shocked Bryarly, not so much for their number but for their condition. Femurs and tibias had been hacked open as if with an ax; skulls had been cut open as if with a meat saw. Whatever sense of spiritual renovation he had felt at the lake quickly faded. Looking at what lay before him, he was suddenly overwhelmed by the sensation that the whole place was, as he wrote in his diary that night, pervaded by a "sad, melancholy stillness . . . which seems to draw you closer and closer. . . ." He grew still and began to reflect on what these bones might have to tell him:

To look upon these sad monuments harrows up every sympathy of the heart & soul, & you almost hold your breath to listen for some mournful sound from these blackened, dismal, funeral piles, telling you of their many sufferings & calling on you for bread, bread.

Wakeman Bryarly had it right. To hear the only voices that can tell this story, you must almost hold your breath to listen. What they have to say is hard to hear, and harder still to come to terms with.

And so, standing in a long-neglected cemetery in the Napa Valley, I, too, held my breath and strained to listen. I pulled a manila envelope from my pocket and took out a photograph of Sarah that had been given to me by one of her relatives. I studied it, as I had done many times before. It has haunted me since I first saw it, and ultimately it is the reason I set out to write this book. At the moment when it was taken, Sarah gazed back at the camera calmly, with bright, intelligent-looking eyes. She was pretty, but only in a quiet, understated way. In the picture her hair is parted in the middle and pulled back above the ears, and she is wearing what looks to be a simple black bodice with a lace collar and a large white bow at the neck. She appears serious, almost somber, but, paradoxically, she is wearing what seems to me to be just the slightest, Mona Lisa–like hint of a smile.

It was those bright eyes and that slim suggestion of an almost-smile that first arrested my attention and that have held it ever since. In an era when people almost universally believed that a grim, even severe, countenance was the most appropriate face to present to the camera, Sarah had let a tiny glimmer of happiness, or at least apparent contentment, show through. Standing there among the rattling thistles next to my great-uncle's grave, I wondered how I was to reconcile that hint of happiness with what I already knew about her life and death.

I turned to go, making my way down through the thistles, around my great-uncle's house, and back to the car. I drove out onto the highway and began an odyssey. To understand Sarah's story, and to tell it well, I knew I needed to go where she had gone, to be where she had been at the time of year when she was there. I needed to go east and then follow Sarah west from where she began to where she ended.

During a series of trips over the next year and a half, I put several thousand miles on various rental cars and stayed at more cheap motels than I care to remember. I waded through waist-high prairie grass crawling with ticks, struggled on foot up dusty inclines in the

Rockies, walked the salt flats of Utah in withering heat, tromped through snow up to my hips in the Sierra, following Sarah's footsteps as closely as I reasonably could wherever she went. In all of this, I saw and felt many things that have enriched my understanding of her story and I hope have enriched my telling of it. Every step of the way, though, it has never been far from my mind that I have not experienced a fraction of the discomfort and hardship that Sarah experienced on any given day during her extraordinary journey toward the ever-declining western sun.

PART ONE

A Sprightly Boy
and a Romping Girl

We look out upon them and are astonished to
see such careless ease and joyousness manifested
in the countenances of almost all—the old, the
young, the strong and feeble—the sprightly boy
and the romping girl. . . .

—*ST. JOSEPH GAZETTE,*
May 8, 1846

1

HOME AND HEART

The night before Sarah left Illinois for California, a full moon—as plump and promising as a pearl—hung over Steuben Township. Down in the bottomlands, the Illinois River slid silently past Franklin Ward Graves's homestead, measuring out time swiftly and irrevocably. On the sliding surface of the water, the moonlight shimmered softly, but beneath the surface the river was black and swollen.

At 5:25 the next morning, April 12, 1846, the sun rose over the bluffs on the eastern side of the river, and, with the sun, Sarah also rose. She climbed from her bed and dressed and made her way down to the house in the bottomlands, the house where she had grown up. In her father's farmyard, teams of broad-beamed oxen stood sullenly in their yokes, their great heads swaying from side to side in the cold, gray light of dawn, white plumes of breath issuing rhythmically from their fleshy nostrils. The oxen were yoked to three narrow farm wagons, each no more than four or five feet wide but nine or ten feet long and covered with canvas stretched on hoops to make a high canopy. Sarah's parents and her eight younger brothers and sisters hurried about in the muddy yard, loading the last few items into the wagons. A few minutes later, the full moon slipped below the bluffs on the western side of the river, an omen, perhaps, of good things to be found in that direction. At any rate, it must have seemed that way.

Sarah was a happy young woman that morning. Just two weeks before, though, things had been very different. She had been deeply torn—in love with a young man who played the violin, but faced with the hardest decision of her young life. More than in love, she had been engaged to become the young man's wife. But her mother and father and siblings were in the final stages of sorting through their things, giving away or discarding many of their possessions and packing the most essential and most precious items carefully into the three farm wagons in their farmyard.

They were about to climb aboard those wagons and disappear over the western horizon, bound for California, a place Sarah could hardly conceive of. If she stayed behind as planned, to marry her young man, in all likelihood she would never see any of them again—not her mother, not her father, not her eight siblings, many of whom she had helped to raise. She could still go with them, of course, but that would mean leaving behind the young man with the violin. Unlike the engagements and marriages of many of her friends out here on the Illinois frontier—some of whom were married off as young as thirteen or fourteen—Sarah's was neither an economic arrangement nor a purely practical matter. Her heart cleaved to the young man's heart.

His name was Jay Fosdick, and he was two years older than Sarah, about twenty-three. He and his family had arrived in Steuben Township more recently than Sarah's family, settling along Senachwine Creek, three miles west of her family's homestead. Like her own family, the Fosdicks were thoroughly Yankee in their origins and their ways. They could, in fact, trace their lineage back to William Brewster of the *Mayflower.* In New England the Fosdicks had for generations been whalers and ships' captains and silversmiths. After the Revolution, many of them had emigrated to western New York and taken up farming. More recently some, including Jay's parents, had come farther west to Illinois in search of better farming country than the rocky soil of New York offered. On arriving in Steuben

Township, Jay's father, Levi, had promptly gone to work planting fruit trees on his new land, creating a minor local marvel that had already come to be called "the Big Orchard" throughout the township.

But Levi Fosdick, with two daughters and only one other, younger, son at home, needed Jay to help him with the relentless cycles of work that a frontier homestead and a large orchard required. He did not, his neighbors would later report, cotton to the idea of his son taking off for a place like California—a place that was more than sixteen hundred miles from the United States as the crow flew and perhaps two thousand miles by wagon road, that could be reached only by passing through Indian territory, and that belonged to a foreign government at any rate. Jay would have to stay in Illinois.

So for weeks, with mounting despair, Sarah had watched her family's preparations as their inevitable departure drew closer. It wasn't only Jay she would have to tear herself away from if she broke off the engagement and chose to follow her family. It was also this place, where she had lived since she was six, the only place she could remember as home.

When she and her family had arrived in 1831, the Illinois River had been near the farthest western reaches of the American frontier. Both the bottomlands along

the river and the wide-open prairies atop the nearby
bluffs were virgin land, and a kind of wonderland for
a barefoot girl of six. In spring the prairie grass up on
the bluffs grew five feet tall, a great green sea in which
a child could hide herself for hours. In the clearings,
prairie chickens strutted and danced, kicking up dust,
inflating the bright orange air sacs on their necks,
booming their deep, low hoots and cackling their shrill,
loud mating cries across the windswept countryside.
Down along the river, the Indian corn in her father's
fields grew even taller than the prairie grass, sometimes
as tall as twelve feet by midsummer. Pumpkins three
feet across grew side by side with bright yellow summer
squash and fat green watermelons. In the spring the
woods below the bluffs trilled with birdsong. In the
dappled light of those woods, Sarah could gather hazel-
nuts by the bushel and fill buckets with sweet, creamy
pawpaws.

And then there was home itself—a one-room log
cabin with a puncheon floor. It was simple and plain,
but by the spring of 1846 every nook and cranny must
have held sacred memories for Sarah. A large clay-
and-wattle fireplace stood at one end of the room. In
the fireplace, nestled among the ashes, a large cast-
iron Dutch oven stood always at the ready. It was a
vessel from which Sarah had likely served and eaten a

thousand meals. For as long as she could remember, it had served for uses as varied as stewing venison, baking salt-rising bread, and soaking her father's frostbitten feet on winter nights.

Trundle beds covered with homemade quilts were tucked into the corners of the room. The family's clothes, almost all of them homespun—woolen stockings, trousers, and skirts; linen dresses, shifts, and shirts—hung from pegs along one side of the room. On the other side, Sarah's father tacked the musky-smelling pelts of wolves and beavers and occasionally bears that he had shot or trapped. Beneath the pelts, gunnysacks full of sweet-smelling wheat and shelled corn lined the wall. A glazed window that admitted sunlight even on frosty winter days looked out on the public road that passed near the cabin. Under this window sat a chest, large enough that Sarah and one of her sisters could curl up on top of it together and pass long hours in a pool of sunlight, sewing or reading or stringing glass beads on threads. Outside the window this spring, as every spring, young chickens scratched and hunted for bugs.

There were neighbors and friends that Sarah would have to leave behind as well if she set out for California. Throughout their little community—the scattered homesteads on the western side of the river and the little village of Lacon on the eastern side—Sarah and her parents were well known and widely regarded with affection. Villagers in Lacon looked forward almost every morning to the sight of her father crossing the river in his homemade canoe.

Franklin Ward Graves cut a striking figure both in his canoe and out of it. He was a tall, lanky man with unkempt hair, and, except in the dead of winter, he seldom wore shoes and never wore a hat. But he was congenial and of a sunny disposition, and when he crossed the river, he brought things the villagers were eager to have—fresh game and cured pelts from his woods, vegetables from his garden plot. They enjoyed his warm and generous nature, his willingness to lend a hand whenever it was needed.

In the afternoon, after Franklin had returned to his side of the river, the villagers were equally pleased when Sarah's mother, Elizabeth, took her turn with the canoe and crossed over to the village bearing honey, eggs, butter, and buckets of soft soap. She, too, was tall and lanky. Spring, summer, and fall, she almost always wore the same blue calico frock, simple cow-skin shoes tied at the tops with leather strings, and an old calico sunbonnet. In winter she made only one concession to the cold—she supplemented her outfit with the addition of a pair of blue woolen stockings. Like her husband, she seemed to wear a constant smile and was always ready to lend a hand.

Years later, when their former neighbors thought back on the Graves family, they remembered them not only as openhearted and generous but also as extraordinarily hardy people. One of those neighbors recalled a particular day during the bitterly cold winter of 1839–40. Elizabeth Graves had come by to deliver some butter, and, finding the neighbor with a newborn infant, she stayed for several hours to help with the household chores. As in every frontier community, the women of Lacon and Sparland lived within a kind of mutual-aid society—a circle of women with whom they could share burdens and confidences in a way that they could not share them with the men in their lives.

As evening drew near and Elizabeth prepared to leave, her neighbor begged her to stay. Elizabeth had her own infant with her, and the woman feared that it would not be safe for the two of them to cross the icy river in the canoe in such cold weather. Elizabeth, as usual, was wearing only her blue frock and blue stockings. But she laughed her neighbor off. She wrapped the baby in a square of linsey-woolsey, declined the offer of a shawl, and set out into the dark, saying she would return with some helpful herbs in the morning. The neighbor watched her lay the baby in the bottom of the canoe and paddle off into the frigid darkness.

That night, a storm raged across the Illinois prairies and a freezing wind howled up the river, shuddering through the chinks and cracks of the log cabins of Lacon on one side of the river and Steuben Township on the other. By the next morning, the river had frozen over, and the neighbor's window was too encrusted by frost to allow her to see anything through it. But at about 10:00 A.M., Elizabeth Graves, still wearing her blue calico frock, came knocking at the door with a handful of herbs. She had walked back across the newly frozen river to fulfill her promise. Referring to the whole Graves family some years later, knowing what had become of them, the neighbor marveled at an irony that was by then all too apparent to her. "They were not delicate hothouse plants," she said.

As Sarah agonized and mourned in advance the loss of either her family on the one hand or her fiancé, home, and neighbors on the other hand, Franklin Graves, at fifty-seven, saw things very differently. Like his neighbor Levi Fosdick, Graves was a New Englander, by birth and by heritage. Born in Vermont's Green Mountains in 1789, he could trace his own Yankee heritage back nearly as far as Fosdick could. In fact, Franklin's father, Zenas, like Levi's father, had served as a fifer in the Revolution. After the war, Zenas had moved his family west to Dearborn, Indiana, one of thousands of his countrymen who yearned for more and better farming opportunities than could be offered by the rocky soils of New England.

Coming of age in Dearborn, Franklin had married a local girl, Elizabeth Cooper. After their first child died in infancy, Sarah was born in January of 1825. Over the next twenty years, eight more children followed. As their brood expanded, Franklin and Elizabeth Graves eventually left Indiana, doing as Franklin's father had done, seeking cheaper land and better opportunities elsewhere. They moved briefly to Mississippi and then on to Marshall County, Illinois, in 1831.

The land across the river from the village of Columbia, later renamed Lacon, had looked good to Franklin Graves. For men like him, a generation of men with rapidly growing families, a surplus of energy, and the desire to convert raw land into an engine of wealth, the quality of a given location's land—the depth and color of its soil—was everything when it came to deciding on a place to settle down. A village of nearly a hundred Sauk Indian wigwams sat in a hollow below the bluff, but the bottomlands appeared to be good wheat country, with deep, black, sweet-smelling soil, so he purchased a tract of the land from the Sauks. He got along well with his native neighbors at first. With so few settlers in the area, there seemed to be enough good land for him and them both, and relations were at least cordial. Recently, in fact, a local chief, Senachwine, had sat by the sickbed of one of the Graveses' neighbors afflicted with a fever, fanning her and caring for her while her husband went in search of a doctor.

Like the few other white settlers just beginning to trickle into the area, Graves scorned the windy and wide-open prairie lands up on the bluffs and instead selected land that was down near the water and timber. It was a formula that had worked for previous generations of Yankee settlers farther east. Aside from the fact that the alluvial soil along the river was richer than the soil up on the prairies, it just seemed to make sense that one would want to be near the two essentials for building and maintaining a home—wood and water. But as it turned out, these particular woods and this particular water added up to a recipe for trouble.

From the time the first white settlers showed up in western Illinois, they had been ravaged by what they called "the ague," or as they soon came to call it, "the Illinois shakes." Come every spring, they would be laid low by high fevers, chills, blinding headaches, stiffness, aching joints, anemia, and, most characteristically, violent and uncontrollable shaking. Usually it would go away after a week or two, but then it would come back a few weeks later, and then again a few weeks or even years after that, and then yet again. It was relentless, and between shivering with the cold in the winter and shaking with the ague in the spring and summer, it wore a man down.

By the late 1830s, the ague was endemic to the whole Mississippi River drainage. Other plagues, ranging from yellow fever to cholera, were also popping up all over the countryside. Steamboat captains would sometimes refuse to stop at some towns for fear of contagion.

The settlers along the Illinois River, like their forefathers in New England and England, did not know what to blame for these outbreaks. They had inherited from those forefathers a general belief that damp weather and stagnant air—"miasmas," as they called the combination—were a principal cause of the fevers. But there wasn't much they could do about either the air or the weather. So they did what they had always done—they dosed themselves with folk remedies. They took pills made from cobwebs. They boiled quarts of water down to pints, "to make it stronger," and then drank it as if it were medicine. They made tea of herbs like boneset (so called because the leaves are joined at the base, leading many to believe that when wrapped around broken bones they would help set the bones as well as drive off fevers). They took patent medicines laced with mercury and suffered far more terribly from the cure than from the cause. They bled themselves dry and wondered why they grew ever weaker and more tired.

The culprit behind the ague was one of the world's great scourges, a nasty little parasite of the genus *Plasmodium*—malaria. Today half the world's population is threatened by malaria, as many as 500 million people are infected each year, and about a million die, twice as many as twenty years ago. Most of the fatalities are children in sub-Saharan Africa, where a particularly vicious strain, *P. falciparum*, predominates. Fortunately for the settlers of Illinois, the two strains of the malaria parasite that made their way from Africa to the American Midwest, traveling in the bloodstreams of infected slaves and sailors, were somewhat less virulent than *P. falciparum*. These two strains—*P. vivax* and *P. malariae*—while less likely to cause sudden death, are more likely to cause chronic problems, cycles of on-again/off-again symptoms that may continue for weeks, months, or even years. It is the relentlessly cyclical nature of the resulting fevers that takes a toll.

What all strains of malaria have in common, of course, is that they are spread to and among humans by mosquitoes, specifically female mosquitoes of the prevalent genus *Anopheles.* Western Illinois, because of the poor drainage that its soils and hard limestone substrata offer, was boggy country in the 1830s and 1840s. Water pooled in the lowlands, and wide fringes of swampy land bordered rivers like the Illinois. All of this provided an ideal habitat for mosquitoes. So when settlers like Franklin and Elizabeth Graves chose to build their homes near water and woods rather than on the open, windswept prairies, they chose to make mosquitoes their close neighbors and constant companions.

When the swarms of insects became unbearable, the settlers built smoky fires in their houses to drive them out, then crawled back into the houses on hands and knees, under the smoke, and closed their doors against the pests. But because the settlers lacked even a basic understanding of germs, let alone the concept of how insects act as vectors for the transmission of those germs, they never thought of the mosquitoes as anything more than highly irritating nuisances.

Bad as they were, the endless cycles of ill health that plagued the settlers along the Illinois River were not the only reason that Franklin Graves was ready to pick up and leave Steuben Township by the spring of 1846. Things had been hard all around. He had bought his land to grow wheat, and grow it he did. The soil was so fertile that the wheat fairly burst out of the ground in the spring, rising up tall and green and rank. By midsummer the stalks were golden and bowed over, the heads heavy with grain. But forces soon came into play that rendered all that bounty more or less useless. These forces worked silently and from a great distance far to the east of Steuben Township, but they were inexorable, and like the ague they ground a man down.

In the spring of 1837, six years after Sarah and her family arrived in Illinois, the United States suddenly found itself at the beginning of a depression so deep that it would eventually rival the Panic of 1893 and the Great Depression of the 1930s. The causes were complex, but they revolved around rampant speculation in real estate. Serious investors and mere speculators had begun to contemplate and then fantasize wildly about the opportunities that new railroads and canals like the recently completed Erie Canal would open up in the West, particularly in the Mississippi River Valley. They bought up land in towns that did not exist, except in their imaginations. They bought into potential railroads that carried phantom passengers from one imaginary town to another. The speculative fever rose and spiraled out of control. The result was the catastrophic collapse, in April and May 1837, of 343 of the nation's 850 banks. In New York alone, commercial establishments lost almost $100 million within weeks, and the effects spread across the nation with devastating results.

As Franklin Graves and his neighbors watched helplessly, the markets for anything they might produce dried up and disappeared. By 1842, wheat that cost Graves fifty cents a bushel to produce fetched a mere twenty-five cents, and then only in the unlikely event that he could find a buyer. If he grew corn instead and used it to raise hogs, he learned that pork that had once brought $4.20 per hundredweight now brought less than a dollar. In some ways Graves and his neighbors were still better off than when they had arrived in this country ten years before, with little more than ambition and what they could carry in a farm wagon. At least they now owned their own land. But many of them were reduced to subsistence-level farming, hunting, and a barter economy, depending on themselves and on one another for nearly all the necessities of life while they watched and waited to see whether things would ever improve.

As they waited, they began to dream again of a better place, a place where the climate was dry and healthful, a place without disease-spawning miasmas, a place free of frostbite and killing blizzards, a place where hard work would yield a crop of ready cash. And in all that, they were about to be obliged. As far back as

1840, the St. Louis newspapers had begun publishing glowing descriptions of a Mexican territory in the far West called Alta California. That same year the novelist Richard Henry Dana Jr. had published *Two Years Before the Mast,* in which he painted a vivid and romantic portrait of a fertile and bounteous California, replete with

> fine forests in the north; the waters filled with fish, and the plains covered with thousands of herds of cattle; blessed with a climate, than which there can be no better in the world; free from all manner of diseases, whether epidemic or endemic; and with a soil in which corn yields from seventy to eighty fold. In the hands of an enterprising people, what a country this might be!

And when a book titled *The Emigrants' Guide to Oregon and California* was published in Cincinnati in 1845, word of the promises it contained spread through the hamlets and homesteads of the Mississippi River Valley like a messianic message.

The author of *The Emigrants' Guide,* Lansford Warren Hastings, was a tall, dashing, and energetic figure, given, when traveling through the West, to dressing as a mountaineer, in fringed buckskin suits trimmed with plucked beaver fur. He was also a lawyer. From Mount Vernon, Ohio, originally, Hastings had traveled extensively through Oregon and California in 1842 and 1843 with the particular goal of sizing up those territories for potential American settlers. In the process he hoped to build a reputation, and perhaps a political career for himself in one of the new lands. A pair of inconvenient facts—that Oregon was at the time jointly and uneasily claimed by both Great Britain and the United States, and that California was sovereign Mexican territory—held little sway for Hastings and the other Americans nosing about in the West.

A number, in fact, had already taken a chance and settled in Oregon. A few had even put down roots in California. Among the latter, an even smaller number had complied with the Mexican government's requirement that they obtain official immigration documentation and become naturalized Mexican citizens. Most simply ignored both the Mexican government and its requirements and thus became California's original illegal immigrants.

What Hastings had to say about California in *The Emigrants' Guide* mesmerized many of the settlers in Illinois and Missouri. He told of the "vast extent of its valleys and plains," the "unexhausted and inexhaustible resources," and "the extraordinary variety and abundance of its productions." He assured his readers of California's "salubrity of climate." And he forestalled concerns about the rights of the Mexican Californians by painting them as little more than savages.

> The higher order of Mexicans, in point of intelligence, are perhaps about equal to the lower class of our citizens. . . . More indomitable ignorance does not prevail among any people who make the least pretensions to civilization; in truth they are scarcely a visible grade, in the scale of intelligence, above the barbarous tribes by whom they are surrounded.

As for the "barbarous tribes," the California Indians, Hastings pointed out that many of them were already conveniently dead, mostly as a result of diseases introduced by the whites.

> For villages of fifty or even a hundred of these huts are frequently seen . . . which are now abandoned, the ground at and all around which is covered with human skulls.

Hastings, in fact, had personally contributed to the mortality in 1843, en route from Oregon to California, when he and the party he was leading had massacred somewhere between twenty and forty Native Americans along the way.*

By the spring of 1846, the American economy had finally pulled out of the doldrums enough to bring some renewed interest in the farmlands along the Illinois River and the products of those lands. So when one George Sparr showed up in Steuben Township that spring looking to buy some land, and with the hard cash to do so, Franklin Graves jumped on the opportunity.

On April 2 he went to the courthouse across the river in Lacon and signed the papers granting Sparr all five hundred acres of his land in exchange for three dollars an acre, fifteen hundred dollars in cash. When he had the money securely in hand, Franklin Graves went home and took an auger and bored a series of holes in a pair of wooden cleats. Then he slipped the greater part of the money, almost all of it in silver coins, into the holes and nailed the cleats under one of his three farm wagons. Looking at them, no one would know that the cleats had any purpose other than supporting the wagon.

---

* One of the deadliest of the plagues that afflicted the California Indians was—ironically enough for the white emigrants who sought to flee "the ague" in the Midwest—a virulent strain of malaria brought from Hawaii by the Hudson's Bay Company in 1833.

The coins within the cleats, a very substantial nest egg for a frontier family of the 1840s, were of various currencies—Mexican pesos, French five-franc pieces, U.S. half-dollars, Spanish dollars, a Bolivian dollar, and a Saxon five-mark piece—some dating back as far as 1806. Neither Franklin Graves nor anyone else on the Illinois frontier was fool enough to take government paper money in exchange for land. With California beckoning and a cache of silver hidden in his wagon, he must have felt as if the whole world were finally going his way.

And if her father had reason to exult on April 2, Sarah had even more. On the day that Franklin Graves went to the courthouse to sign away his land, Sarah went with him. So did Jay Fosdick. Together they stood before Justice of the Peace David Dickinson and were married. They had resolved to go to California together, to build their new life there. Mrs. Jay Fosdick, as she could now call herself, had all she could hope for: her young man, her beloved family, the prospect of a home of her own in a place said by many to be a sort of paradise on earth, and an opportunity to see a world far wider and far more exotic than the narrow confines of the damp river bottoms in which she had grown up.

Sixteen hundred miles to the west, though, something else was afoot that would soon affect Sarah and Jay in unfathomable ways. Lansford Hastings had been talking to John Sutter—a Swiss immigrant who more than matched Hastings for outsized ambitions and a sometimes casual regard for the truth.

Since arriving in California in the summer of 1839 with a retinue of Hawaiian laborers, Sutter had begun to unfold a grand plan in the Sacramento Valley— first to build a substantial fort as a defense against the local Indians and second to found a colony called New Helvetia. The colony, he hoped, would be populated by future waves of Swiss immigrants—artisans and farmers who would produce a wide variety of products to enrich the settlement and its proprietor.

Over the past several years, Sutter had systematically brought the local Miwok Indians under his control, blending diplomacy and generosity on the one hand with brutal discipline on the other. Hundreds of Miwoks had been engaged in building thick, high adobe walls around the fort. A select few had been enlisted into Sutter's homemade personal army, dressed in green-and-blue Russian uniforms and set to drilling in the central courtyard of the fort.

From time to time, when factions among the neighboring Indians had grown too bold and threatened to steal his horses, Sutter had meted out violent, preemptive justice. In a dawn raid on a nearby encampment of Miwoks in 1840, Sutter and his men had opened fire with cannons and small arms, killing about thirty people without warning. For minor transgressions he relied mostly on imprisonment and whippings, though he also sometimes ordered executions. An admiring visitor, William Wiggins, commented that Sutter was "the best Indian tamer and civilizer that I know of."

Now Sutter and Hastings had been contemplating what Sarah and thousands like her would mean to them and their personal fortunes. Hastings had been bending Sutter's ear, telling him that thousands of American emigrants, perhaps twenty thousand or more, were on their way to California, largely as a result of his book. Many of them, he said, were Yankee farmers, men of substantial means, men with capital to invest in California real estate, a commodity that Sutter possessed in vast quantities.

What Lansford Hastings had to say struck a chord with John Sutter. By early April, Hastings had finished helping Sutter lay out the site of a future town—Suttersville—on a patch of Sutter's land located on a bluff three miles southeast of the fort. In exchange for Hastings's help, some trade goods, and additional unspecified services, Sutter had promised to build Hastings a large home and a general store and to give him title to a portion of the lots in the new metropolis.

First, though, Hastings had something urgent to take care of. In *The Emigrants' Guide*, while he had made California out to be virtually a land of milk and honey, he had also sung the praises of Oregon, a destination that many of the emigrants felt was more easily and more safely reachable than California. He had written his paean to Oregon, however, before he had worked out his arrangement with Sutter. Emigrants who went to Oregon would not be interested in driving the Mexicans out of California, nor in purchasing lots in Suttersville. Hastings needed to divert them.

Luckily, from his point of view, in *The Emigrants' Guide*, he had inadvertently given the California-bound emigrants an extra incentive to head his way—a shortcut to California, a road that would shave several hundred miles off the trip.

> The most direct route, for the California emigrants, would be to leave the Oregon route, about two hundred miles east from Fort Hall; thence bearing southwest, to the Salt Lake; and thence continuing down to the bay of St. Francisco, by the route just described.

At the time he had written about it, the mention of the potential shortcut probably hadn't meant much to Hastings. At some point while composing his book, he had evidently contemplated one of the vaguely drawn maps of the American West then in circulation and noted that the current route to California looped inefficiently to the northwest to reach Fort Hall in present-day Idaho, before dropping down to the latitude of what we now call Donner Pass, where it entered California. Why not, he must have thought, go southwest from what is now western Wyoming to the south end of the Great Salt Lake and then more or less directly west to intercept the established route to California? It must have seemed a self-evident improvement to the Fort Hall road.

The problem—and it was a substantial one, as he was just beginning to realize—was that Hastings had never taken his own shortcut, which ran directly through the Wasatch Mountains. In fact, except for a few trappers and a mounted expedition under the command of John C. Frémont and Lieutenant Theodore Talbot the previous fall, no one had ever attempted the route, certainly no one riding farm wagons laden with twenty-five hundred pounds of goods drawn by teams of plodding oxen. Anyone who had done so would not likely have suggested it to anyone else. But Lansford Hastings had made many promises in California, and, for him at least, the shortcut offered an opportunity to seal his reputation as a trailblazer, his potential leadership role in a California free from Mexican control, and now his immediate financial interests as well.

And so on April 11, 1846, Lansford Hastings was already in the saddle in California's Sacramento Valley. He was riding east through green foothills toward the still snowcapped Sierra Nevada Mountains, setting out to meet Sarah and what he expected to be vast legions of emigrants like her, to show them the way to Suttersville.*

The next morning, April 12, Sarah took one last, long look at the place that had almost always been her home. Then Jay flicked a switch at the rump of one of the oxen and gave a shout. The oxen leaned into their yokes, and the wheels of the wagon turned—the first slow revolution of thousands to come—and Sarah and her family, walking alongside their wagons, began to climb out of the cold, wet bottomlands along the Illinois River, one footstep at a time.

---

* Hastings's imagination and his rhetoric had vastly outstripped the reality of the 1846 emigration. However, it should be noted that his expectations were no doubt influenced by his dealings with Sam Brannan, who was that spring actively working to bring more than ten thousand Mormons west to California. On March 26, Hastings wrote to John Marsh, an American settler with a large rancho on the western side of the Sacramento Valley, "You can rely on . . . six or seven thousand human souls . . . [but] the emigration will be much more likely to amount to fifteen or twenty thousand." There were, in fact, just a few more than two thousand California-bound emigrants on the trail that spring and summer.

MUD AND MERCHANDISE

In 1846 spring came slowly to the Illinois River Valley. All through March and the first half of April, an iron-gray sky sat over the bottomlands like the close-fitting lid on a Dutch oven. Day after day relentless rain and snow and sleet slanted down out of the opaque heavens. The countryside was boggy, the Illinois River swollen and spilling over its banks.

The Graves family's initial objective was to get to St. Joseph, Missouri, where they could count on purchasing all the supplies they would need for the long overland trip and where, equally important, they could count on meeting up with other emigrants to form a traveling party. St. Joe, as it was just then beginning to be called, had the advantage over the other principal jumping-off place—Independence, Missouri—of being sixty miles closer to the Great Platte River Road, the established route to the South Pass through the Rocky Mountains. St. Joe lay to their southwest, but they began by going northwest to a tiny hamlet called New Boston, where a ferry transported their wagons across the Mississippi River. Then they began to travel west and southwest, cutting across southern Iowa and slanting down through northern Missouri.

It was slow going. The whole country seemed to be underwater. Six days after they left, the *Illinois Gazette*, back home in Lacon, noted that a particularly long siege of rain had finally ended, but it also noted a most unusual event—the Illinois River now was high enough "to float steamers of the largest class." With the rain, every stream or low spot Sarah and her family came across presented an obstacle that would have to be overcome. Wagons had to be double-teamed to get through boggy spots, small creeks had to be bridged or scouted for places where they could be forded, larger streams had to be searched for a ferryman willing to hazard the crossing, and with a ferry big enough to accomplish it. If a ferry could not be found, the only way across was to build a makeshift ferry of their own, or swim the livestock across and then caulk the beds of the wagons and use them as rafts, a hazardous and arduous endeavor at best.

Mud was their constant companion—it squelched under the heavy feet of the oxen, it plastered the withers of their saddle horses, it flew out from under the turning wagon wheels, it splattered their clothes and their hair and their faces. They scraped mud from their boots, they daubed it from their eyes, they combed it out of their hair, they dug it out from under their fingernails, they tasted it in their food, and they cursed it all the while.

There were thirteen of them, divided among three wagons: Franklin and Elizabeth likely drove the wagon containing their household goods and the hidden coins. Jay and Sarah, a household unto themselves now, drove a second wagon. A young man named John Snyder who had moved to Steuben Township from Ohio the previous winter drove the third. Snyder, at twenty-three, was muscular, strikingly handsome, and notably genial. He had a buoyant, carefree way about him that put others at ease. Hearing that the Graves family was bound for California, he had asked if he might travel with them. Franklin Graves, at fifty-seven, knew that he could use the muscle power of another young adult male, so he struck a deal with Snyder—he could drive the third wagon and perform other chores in exchange for his board until they reached California.

Sarah and her many siblings traveled variously in the different wagons, on foot, or on saddle horses. Mostly they walked, to spare the horses and the oxen on whose strength and endurance their success would eventually depend. Sarah's closest sister, nineteen-year-old Mary Ann, was widely reputed to be a notable beauty, with dark, wavy hair and a broad white smile. Her oldest brother, Billy, seventeen, was a gangly teenager, built much like his father, already pushing past six feet in height. The younger children were Eleanor, thirteen; Lovina, eleven; Nancy, who would turn eight on April 26; Jonathan, seven; Franklin Ward Jr., five; and the baby of the family, Elizabeth, only about nine months old.

The first green of the spring prairie grass had begun to emerge from beneath the old, dry, butterscotch-colored grass of the previous year. As they drove over the wide-open countryside of southern Iowa, the dead grass whispered continuously around their wagons, stirred by the ceaseless prairie wind. Other than that and the muffled footfall of the oxen on the wet earth, they moved through a largely silent world.

As they turned south into Missouri, the terrain began to change, the amplitude of the hills—the distance from the lowest points of the bottoms to the crests of the hills—began to diminish, and the intervals between the crest of one hill and the crest of the next began to increase, the land still far from flat, but moving in that direction. The weather grew mixed, sometimes wet and sometimes fair, but the cold gradually began to moderate, and the wet wind began to carry hints of spring. Sarah and her siblings were in high spirits. As they walked or rode alongside the wagons, they spent hours imagining and chatting about the wonders that lay before them in California and the adventures they would have on the way there.

None of her siblings' hearts, though, could have been as light as Sarah's. In the 1840s it was customary for fashionable brides to take some form of what was variously termed a "wedding journey" or "bridal tour," often in the company of friends or family. The word "honeymoon" had been in use long before the 1840s but referred more generally to a period of presumed marital bliss following the nuptials rather than to a journey. The brides of successful New Yorkers might expect to make a wedding journey upstate to Niagara Falls. Brahmin girls from Boston's Beacon Hill might make extended tours of Europe. Country brides like Sarah, though, were seldom able to afford any sort of celebratory journey at all. If she and Jay had stayed in Steuben Township, the most she might reasonably have hoped for was a twenty-mile trip down the river to spend a night or two in Peoria. But instead here she was, embarking on an epic journey to a fabled land. Ahead of her lay not only the prospect of a home and a farmstead of her own in a rich land with a benign climate but also the opportunity to see and experience things she had only read of or imagined before now— surf pounding on a sandy beach, the smell of salt air, ocean fog filtering through dark forests, odd beasts such as antelope and sea lions, dusky-skinned "Spaniards" (as the emigrants often called Mexican Californians), and magnificent mountains. Mountains of any kind would be a novelty, but Sarah knew that ahead of her lay mountains that were reputed to rival the Swiss Alps, picturesque peaks capped by granite crags and draped with deep drifts of snow.

And so they moved slowly and blithely toward St. Joe. But even as they contemplated the prosperity that lay ahead of them, people were in motion and events were unfolding in faraway places—people and events that were as yet unknown to them but that in time would come together to profoundly alter the world they were about to enter and ensnare them in a deadly web.

On May 12 a party of emigrants bound for California departed from Independence, Missouri, south of St. Joe. One of the organizers of the party was forty-

five-year-old James Frazier Reed, a businessman from
Springfield, Illinois. Reed was relatively affluent and,
according to many who were with him that spring,
rather full of himself. Though not officially the captain of the group, he seems to have regarded himself
as its natural leader from the outset. Traveling with
him was his thirty-two-year-old wife, Margret, who
suffered greatly from migraines, or "sick headaches"
as they were then called, and was generally frail—so
frail, in fact, that she had lain in bed at her wedding,
with her husband standing by holding her hand. The
couple hoped the climate in California would cure her.
The Reeds had in tow their children and stepchildren;
Margret Reed's seventy-year-old mother, Sarah Keyes;
a cook; a personal servant; and several teamsters whom
Reed had hired to drive wagons and handle his livestock.

Anticipating the rigors of the journey ahead, James
Reed had gone to extra lengths to prepare at least one
wagon that would have in it some of the amenities of
home for his wife, children, and aging mother-in-law.
His stepdaughter, Virginia, described it in considerable detail.

The entrance was on the side, like that of an old-
fashioned stagecoach, and one stepped into a small

room, as it were, in the center of the wagon. At the right and left were spring seats with comfortable high backs, where one could sit and ride with as much ease as on the seats of a Concord coach. In this room was placed a tiny sheet-iron stove, whose pipe, running through the top of the wagon, was prevented by a circle of tin from setting fire to the canvas cover. A board about a foot wide extended over the wheels on either side the full length of the wagon, thus forming the foundation for a large and roomy second story in which were placed our beds. . . .

It was, perhaps, not quite the ponderous "palace car" that later mythology would make it out to be, but apparently it was substantially more elaborate and comfortable than the simple, roughly four-by-nine-foot farm wagons that most of the emigrants drove.

Traveling with the Reeds were two brothers, also from the Springfield area—George and Jacob Donner—and their families. George Donner, in his early sixties, was among the older emigrants setting out for California that spring. A prosperous farmer, he was comfortable in his own circumstances, but he hoped to relocate his five daughters to a place where they would find more favorable prospects than in Illinois. Like so many of the emigrants of 1846, he was a large man physically and the son of a Revolutionary War veteran. He had been married three times before. His present wife, Tamzene, was at forty-four a small woman with gray-blue eyes, dark hair, and, according to one source, a "not pretty" face. But she was an accomplished woman. She had twice been a schoolteacher, she spoke French, and she was an enthusiastic amateur botanist. She also had been married before, in North Carolina, but in the course of less than a year she had lost her entire family—a daughter, born prematurely, as well as her husband and her son to illness. Jacob Donner, George's brother, thought to be about fifty-six, was a slight man and not in robust health. He and his wife, Elizabeth, forty-five, had between them seven children. Like the Reeds, the Donner brothers and their families had brought along a number of hired teamsters to handle the livestock and drive two of their extra wagons.

Events that would affect Sarah and her family, the Reeds, the Donners, and everyone else setting out for California that spring were also unfolding far to the east, in Washington, D.C. On May 13, President James K. Polk signed a bill declaring that a state of war existed between the United States and Mexico.

Polk had wanted this war—or at least the booty it would yield—from his first days in office. Though the war was ostensibly fought over a border clash between Mexican troops and U.S. troops in the newly independent but disputed Republic of Texas, the real prize was California. California at the time included not only the modern state we know today but also all of Nevada, Arizona, and New Mexico, as well as parts of Colorado and Wyoming—nearly half of Mexico's territory at the time. Polk had earlier tried to buy California on the cheap, but that having failed, he had resolved simply to take it. The dispute over Texas had given him the opening he wanted, and it dovetailed nicely with a carefully choreographed campaign of presidential deception that would find a disconcerting parallel early in the twenty-first century.

Many in the Congress, and some in Polk's own cabinet, thought that the outright seizure of California would be unjustified and immoral. But Polk was determined to defend and expand his executive powers against any congressional interference. He was also at times stubborn and narrow-minded. Historian Bernard DeVoto said of him that he was "pompous, suspicious, and secretive; he had no humor; he could be vindictive; and he saw spooks and villains." And apparently he counted anyone who disagreed with him about Mexico and California as primary among the spooks and villains. When he sent his war bill to Congress on

May 11 and the members of Congress decided that perhaps they should debate the issue before acting on it, Polk was outraged, taking their failure to act immediately to be nothing less than treasonous.

After months of jingoistic rhetoric emanating from Washington about the natural right of Americans to fulfill their manifest destiny and the obvious moral depravity of Mexico and Mexicans, the mood of the country was largely behind Polk. The urge to expand the nation's territory was almost universal, and it seemed self-evident to most Americans that they had a natural right to as much of the continent as they desired. The administration had crafted its rhetoric carefully, advertising the impending conflict as a defensive war, not as the war of conquest that it in fact would be.

Faced with this kind of popular support, the Congress passed the war bill 42–2. At a cabinet meeting that evening, Polk's secretary of state, James Buchanan, still had his doubts. He suggested that perhaps the administration should make clear to foreign governments that the United States' argument with Mexico concerned only Texas and that the United States had no designs on California, New Mexico, or any other part of Mexico's territory in the West. The president's angry reaction was to proclaim that certainly he would take California if he could, that he would "acquire California or any other part of Mexican territory which we desire."

On May 16 the headline for the *Illinois Gazette* back in Lacon read "WAR! WAR!" Two days later the Republican candidate for Congress in the Seventh District of Illinois, Abraham Lincoln, arrived unannounced in Lacon, campaigning for the congressional seat he would soon use to denounce the war with Mexico.

On May 19 the Donner and Reed families, having crossed both the Missouri and the Kansas rivers, fell in with a much larger party led by a Kentuckian, Colonel William Russell, near a Kaw Indian settlement known as Fool Chief's Village. The Russell Party, glad to have the addition of more men—particularly well-regarded, relatively educated, and affluent men like James Reed and the Donner brothers—voted them into their party unanimously.

Out in the middle of the desert sage-lands of present-day Nevada, Lansford Hastings was also in motion. On May 20 he had arrived at a crucial juncture in his mission to intercept the westbound emigrants and direct them toward Suttersville. Since leaving the western foothills of the Sierra Nevada, he had struggled with horses and pack mules, floundering through deep drifts of snow while crossing the mountains, then making his way down through the basalt canyons of the eastern foothills and out onto the arid plains of western Nevada. By happenstance he was traveling in the company of James Clyman, among the greatest of the generation of mountain men who traveled the American West between the 1820s and the 1850s. Born in 1792 on a farm belonging to George Washington, whom he had seen in the flesh, Clyman had traveled widely in the far West as early as 1823, and it's unlikely that anyone within a thousand miles knew the lay of the land better than he did.

Clyman and Hastings camped that night on the spot where the established emigrant route northeast to Fort Laramie intersected the theoretical shortcut that Hastings had promoted in his book. Now Hastings wanted to take that shortcut, to see it himself for the first time and to reach the emigrants before they could have a chance to take the road to Oregon rather than California. Clyman thought that the proposed shortcut saved little distance and promised much harder traveling than the proven route. They sat by a campfire in the sagebrush that night and argued about it. In the morning they argued some more. But Hastings would not be deterred, and so they left the established road and struck out across the desert. Ahead of them lay the

searing salt flats of Utah and the almost entirely unex-
plored tangle of mountains and canyons known as the
Wasatch.

Crossing Iowa and Missouri, Sarah and her family
fell into a basic and comfortable routine that varied little from day to day. They arose early and built campfires. Franklin and Billy Graves, Jay Fosdick, and John
Snyder rounded up the family's loose livestock and
yoked up the teams. Sarah and her mother and Mary
Ann took out long knives and cut thick slabs of bacon
from the hams hanging ponderously in their wagons
and fried them up in cast-iron skillets. They brewed
strong coffee or tea, and then they sat on the grass or
on wagon gates eating the bacon, sopping up the grease
with pieces of bread they had baked the night before.

Then they threw the last few items into the wagons
and got under way. At midday they stopped for an hour
or two—"nooning," as they called it—to rest the teams,
let the cattle graze, and take another quick meal themselves. Then they continued on again until four or five
in the afternoon. When they found a good spot near
clean water and with ample grass for the livestock, they
combined their wagons with those of other families
and drew them into a square or a circle and turned the
animals out to graze while they prepared campfires,
prepared dinner, set up tents, ate, and gathered around

the fires to socialize a bit before turning in. If they felt that Indians were about, they drove the livestock into the enclosure formed by the wagons and set guards out for the night.

When it came time to bed down, they crawled into the backs of wagons or into tents. We do not know where Sarah and Jay slept. Tents were expensive items, and it was common among the emigrants for whole extended families, meaning children, seniors, men, women, and couples—including newlyweds—to share a single large tent. In chilly nights on the plains, and later in the mountains of the West, it was the simplest and most effective way for everyone to stay reasonably warm. But Sarah and Jay might well have chosen to sleep in the relative privacy afforded by the farm wagon they drove.

In the middle of May, they approached St. Joe. They traveled now alternately through open prairies on the uplands and through virgin woods in the valleys. This was rich, fecund land, country that an aging John James Audubon had prowled just three years earlier, making observations and collecting specimens for his *Viviparous Quadrupeds of North America*, a follow-up to *Birds of America*. The ancient, massive trees of the woods had leafed out, and as the Graves family descended into the Missouri River Valley, they entered a world that was a deep, dense, and somber green, but explosive with flashes of brighter colors. Sky blue irises rose in ranks out of the mossy ground. Scarlet tanagers, orange orioles, and yellow warblers drifted through the canopies of tulip poplar and black locust trees. Deer and gray wolves slipped away silently into the understory. Passenger pigeons, with their slate blue backs and wine red breasts, sometimes nearly filled the sky overhead and at other times draped the enormous trees, weighing down the limbs. Along the river, white pelicans plunged out of the sky Icarus-like, piercing the water with hardly a ripple. Blue-winged teals, green-winged teals, and cinnamon teals gabbled and bobbed in the water with hosts of other ducks. Red-winged and yellow-headed blackbirds clung to reeds along the shores, chortling and chattering among themselves and scolding passersby.

As they emerged from the woods and made their way into the rough-and-tumble town that stretched along a big bend in the river, Sarah and Jay found themselves part of a small flood of emigrants converging on St. Joe that spring. The local newspaper, the *Gazette*, marveled at the number of wagons that appeared and at the attitude their occupants displayed.

From every quarter they seem to come—prepared and unprepared to meet every emergency. We look out upon them and are astonished to see such careless ease and joyousness manifested in the countenances of almost all—the old, the young, the strong and feeble—the sprightly boy and the romping girl, all plod along as if the jaunt were only for a few miles instead of a thousand—as if a week's troubles were to terminate their vexations and annoyances forever. What an idea it gives us, and what an insight into human nature—HOPE, the bright, beaming star is ascendant in their sky, alluring them on. . . .

St. Joe that spring was a vortex of commercial activity—a swirling confusion of horse traders, wagon builders, ferrymen, outfitters, Indians, blacksmiths, gunsmiths, and entrepreneurs of every stripe. Steamboats trailing black clouds of smoke clawed their way up the notoriously shallow and therefore dangerous Missouri River and tied up along the waterfront. Slaves labored to unload the boats, carrying bales and barrels of goods on bare, sweat-slicked backs. Along the riverside on the south end of town, more slaves tended fields of hemp, a crop grown not for its intoxicating qualities but because it was the source of valuable fibers. Flat-bottomed scows, pressed into service as ferries, struggled back and forth across the river carrying white-topped emigrant wagons, the men at their long sweeps fighting the current all the way. Pottawatomie and Kaw Indians walked bare-chested through the streets hawking vegetables and fish and game. Trappers and mountain men dressed in buckskin breeches, fringed jackets, and bear or coonskin hats mingled with emigrants in broad-brimmed hats and sunbonnets, dispensing advice and tall tales in exchange for liquor and tobacco. Stern-eyed Christian missionaries browsed the mercantile stores for cheap trade goods that might appeal to the heathen out west. Merchants come upriver from St. Louis inspected stacks of furs and made cash-on-the-barrel head offers. Brass whistles on the steamboats shrieked. Hammers and anvils in the blacksmith shops rang out rhythmically in point and counterpoint. Oxen bellowed. Dogs barked. Hogs squealed as they trotted toward the slaughterhouse and their doom. The river stank of raw sewage and pig offal. The streets reeked of manure and horse piss. But bakeries also perfumed the sour air with the aroma of fresh-baked bread.

For many of the emigrants, St. Joe was a last chance to see a doctor before leaving the United States behind, a vitally important consideration for people who soon might find themselves a thousand miles from the nearest doctor. According to the local "card of rates" agreed upon by the physicians of St. Joe the previous year, one could have an abscess opened for fifty cents.

For a dollar one could obtain medical advice or receive an enema. A troublesome tooth could be extracted for fifty cents, troublesome toes or fingers amputated for five dollars each, arms for ten dollars, legs for twenty. If an emigrant woman were to find herself in a delicate condition, she might choose to linger in St. Joe long enough to have her baby professionally delivered for five dollars, but there would be no volume discounts—twins would set her back ten.

In the midst of all this activity, Sarah and Jay faced the task of making sure that they were well provisioned for the journey ahead. Some of what they needed they had brought from home, but this was the last and best chance to stock up on any remaining essentials. Selecting the right items and the right quantities was critical to their success—more critical, as it would turn out, than they could yet begin to imagine. Too much in the way of food and gear would weigh them down and tire their oxen, possibly even kill the oxen when the going got tough. Too little would raise the possibility of hunger or even starvation for the family if anything went wrong along the way. The guidebooks, including Lansford Hastings's *Emigrants' Guide*, gave specific recommendations for the quantity of certain staples that each adult traveler ought to procure: "at least two hundred pounds of flour, or meal; one hundred and fifty pounds of bacon; ten pounds of coffee; twenty pounds of sugar; and ten pounds of salt. . . ."

But there were many hard decisions to be made beyond figuring out the required quantities of staples, and for the first time in her life, as the mistress of her own household, it was largely Sarah's responsibility to make these decisions. For one thing, as all frontier women knew, not all flour was created equal.* Sarah had to choose among three basic types that had long been known as "shorts," "middlings," and "superfine." Shorts contained mostly bran and very little of the actual endosperm, the white, starchy portion of the wheat kernel. It was coarse, gritty stuff, often contaminated by dirt, wheat chaff, and insects. Middlings were hardly any better, a mixture of bran and wheat germ, often blended with cornmeal or rye to stretch it. Middlings generally needed further refining to be useful in baking. Superfine flour was stone-ground and passed through sieves. It resembled what we know as whole wheat flour, but it was expensive and beyond the reach of most emigrant families.

Sugar raised another set of issues. It could take the form of molasses in kegs or barrels; sticky, hat-shaped

---

* None of the flour available to Sarah was what you and I know of as white flour. The process for chemically bleaching flour was not commercially available until after 1900.

loaves of brown-and-white sugar; lumps of gooey unrefined brown sugar; or "Havana," a lumpy crushed white sugar that required still more crushing and sifting to be useful in baking cakes or pastries. Because they were relatively inexpensive and portable, the hat-shaped loaves were what most of the emigrant cooks carried, hanging them by strings within their wagons. Most also carried loose-leaf tea and coffee, the latter in the form of green beans that would have to be roasted in a pan over a campfire and then ground in a portable coffee grinder before it could be brewed.

To raise their bread, they took "saleratus," as they called baking powder, a commodity that they would later, to their delight, find occurring naturally around soda springs in the West, where they could scoop up as much of it as they wanted for free. For the many days when they would not have time or inclination to bake bread, they took hardtack or crackers. Most took a tub of clear suet to substitute for butter, though those who brought along a milk cow could simply hang a bucket of cream inside a wagon and let the bouncing and jouncing of the wagon churn fresh butter for them. Almost everyone brought vinegar, which was useful not only to lend flavor to their food but also as a cleaning agent and, they believed, as a medicine for both their own maladies and those of their livestock. Most also brought whiskey or brandy for the same purposes, and for celebrating special occasions like the Fourth of July.

They brought hard candy, hard cheeses, figs, raisins, flavored syrups—lemon and peppermint being particular favorites—salted codfish, pickled herring, and jellies, jams, and preserves packed in stoneware crocks. Some of the items they crammed into their provision boxes carried brand names that you or I might still find on our own kitchen shelves—Underwood's deviled ham for one, and Baker's chocolate with which to flavor a sweet cake or make hot chocolate at the campfire.

Because they came from a homestead, Sarah and her family probably brought along sides of bacon from their own hogs back home, but if not, they could buy as much smoked, salted bacon as they wanted in St. Joe. Except for whatever game they might kill along the way or as many of their cattle as they might choose to slaughter, the bacon would be the only meat they could count on until they reached California.

Because they did not yet know in whose company they would be traveling, nor how events might separate them from others, every family needed to be capable of basic self-sufficiency all the way to California. This meant that if they hadn't brought them from home, before leaving St. Joe each family needed to procure all the requisite tools for maintaining and repairing their wagons, preparing their meals, tending to their livestock, providing shelter at night and during storms, crossing flooded streams, and defending themselves from the Indian attacks that they viewed as exceedingly likely and that they feared above nearly all else.* As a result, despite the burden it placed on their mules and oxen and the compromises it imposed on them in terms of what else they could bring along, they had to carry a great many very heavy items—hammers, chisels, augers, axes, bolts, screws, shovels, tents, frying pans, Dutch ovens, coffeepots, additional cooking utensils, bullets or the lead for making them, iron shoes for their oxen and horses, kegs of black powder, and large numbers of guns.

The guns were of two basic types—older flintlocks, which depended on a hammer striking a piece of flint to create a spark that, with luck, ignited a charge of black powder, or the more recent percussion guns. The latter, in place of the flint, depended on a small copper capsule called a "cap." The cap contained a small amount of highly explosive fulminate of mercury painted on the inside of one end and covered with a drop of varnish. When struck and crushed by the hammer, the cap exploded, igniting a charge of powder

---

* In fact, though, only four emigrants would die at the hands of Indians in 1846, while twenty Indians would die at the hands of emigrants.

and thus discharging the weapon. By 1846 most newer guns, both muskets and pistols, were percussion weapons, which were considerably less likely to misfire as a result of damp powder, faulty flints, or wayward sparks. But many of the emigrants, including Franklin Graves, held to the old ways. Along with their percussion guns, they carried flintlocks that dated back to the American Revolution or earlier, guns their fathers and grandfathers had handed down to them and with which they had hunted game ranging from squirrels to bears all their lives.

By the time Sarah and Jay had fully stocked the wagon that Jay drove for Franklin, it likely weighed as much as three thousand pounds. It was a heavy load, and they knew that it would take a toll on their oxen, but with the things they carried, they believed they were well prepared for what they expected would be at most five or six months on the trail. By the time the first frosts settled on the Sacramento Valley—if such a thing as frost even existed in California—they would be building a new home among golden hills.

There was one other commodity that the emigrants passing through St. Joe that spring tried to stock up on. Advice. Even the best of the guidebooks, they knew, could not substitute for the firsthand and timely knowledge of those who had recently been to the West and returned. The trappers and traders who roamed the streets of St. Joe had detailed knowledge of what lay ahead—the best routes, the best places to ford a river, which of the natives to befriend and which to be wary of, when to expect what kind of weather, how earlier parties had fared—and so the emigrants listened eagerly to what they had to say.

One bit of news that was just working its way back across the plains that May concerned a large party of emigrants who had left St. Joe the previous spring, bound for Oregon.

They had left St. Joe provisioned much as Sarah and her family now were. All had gone well until, in modern-day Idaho, they began to hear rumors that the Walla Walla Indians, through whose territory the road to Oregon's Willamette Valley lay, were hostile to whites and that they should be prepared for a fight ahead. On August 24 a trapper and guide named Stephen Meek overtook them. Having just escorted a train of emigrants from St. Louis to Fort Hall, and traveling now with his recent bride, Elizabeth, Meek was anxious to find new employment. He had traveled through Oregon a number of times before, including an 1843 expedition on which he had briefly served as a guide for none other than Lansford Hastings. On hearing the emigrants' concerns about the Walla Wallas, he proposed to guide them on a shortcut that would steer them safely to the south of any Indian trouble. He assured them that the shortcut would follow old trappers' trails that he knew well and that it would save them perhaps 150 miles off the old road to boot, delivering them to a settlement at The Dalles on the Columbia River. From there they could travel by water down to the mouth of the Willamette River. And he'd do it, he said, for five dollars per wagon.

Meek, while not well known himself, was the brother of a renowned trapper, Joseph Meek. There seemed no particular reason to doubt his knowledge of the country. So over the course of the next few days, August 25–27, the party split up. Most continued on the old road, but nearly two hundred wagons, four thousand head of livestock, and nearly a thousand emigrants—turned off the main Oregon Trail and followed Meek up the Malheur River into the Blue Mountains. The river had been given its name by French trappers after Indians stole some of their beaver pelts, *malheur* meaning literally "the bad hour," but more generally "misfortune." The Meek Party was about to find that the trappers had named the river aptly.

At first the route seemed tolerable, but within a few days things began to go bad. The country was dry and dusty, and grass for the livestock began to grow sparse. By August 30 one of the emigrants, Samuel Parker—whose wife and child would die as a result of what was about to happen—was already growing disgusted. In his journal he wrote, "Rock all day—pore grass—more swaring then you ever heard." When they got to higher terrain, up on the bluffs above the river, they found less dust, but the ground was littered with particularly hard, sharp, angular rocks that lacerated the feet of their oxen and horses. Within a few days, the livestock were marking the trail with blood dried to black slicks on the hot rocks. Soon some of the oxen began to give out. Each day a few more lay down in the road and died. By the seventh day, nearly everyone was complaining about Meek and his cutoff. The terrain grew even rougher, the rocks larger. Wagons began to break down, necessitating long delays to make repairs. By now the party was strung out so far along the trail that Meek was a day or more ahead of some of them. He began leaving notes telling them where they were and where to go next, but the notes soon became confused and contradictory, and it slowly dawned on all of them that Meek was entirely lost. On the tenth day out, one of the party, John Herren, wrote, "We cannot get along fast, and we are rather doubtful that our pilot is lost. . . . Some talk of stoning, and others say hang him."

Then things got even worse. They emerged from the Blue Mountains, crossed over the Stinking Water Mountains, and came out into the bleak high-desert country of central Oregon. Water began to become an issue, and the midday heat was now searing. On the fourteenth day, they made it to a large, shallow lake, Lake Malheur, only to find that its water was alkaline, foul-smelling, and undrinkable. They wandered on across the sterile desert for days, often traveling both day and night now, having to settle for whatever water was offered up by the occasional meager and muddy spring. Even when they found sufficient water for all the people, there often wasn't enough for the livestock. Food began to run out in many of the wagons, and the emigrants were reduced to eating what they called "poor beef," the emaciated and sometimes rancid flesh of their dead oxen.

Living conditions deteriorated rapidly, sanitation suffered, and soon a ravaging fever began to spread through the camps. A woman named Sarah Chambers had succumbed to it just a few days out. Now children began to die, then increasing numbers of adults. Soon the evening burial rituals varied only in the number of corpses interred, sometimes just one or two, other times as many as six. Each time, though, on the following morning another family faced the brutal necessity of turning their backs and walking away, leaving the body of a loved one behind in a lonely grave in a desolate landscape to which they knew they would never return. It was an experience that was all too common on the overland trail that summer and in the summers to come. For some the pain and sorrow were eased by the deep Christian faith that many of the emigrants held. But not all of them were religious by any means, and even those who were often found themselves haunted years later by the memory of those forlorn graves scratched out of gravelly soil under a pine tree and then immediately abandoned forever.

Some of the survivors of the Meek Party later referred to the malady that afflicted them as "camp fever," which suggests that the culprit was an epidemic form of typhus caused by a species of *Rickettsia* bacteria. Transmitted primarily by lice and fleas, epidemic typhus is typically found where large numbers of people are living in unsanitary conditions, as in ships and military and refugee camps. In the absence of treatment with antibiotics, the mortality rate for epidemic typhus runs as high as 60 percent. Death is generally preceded by headaches, high fevers, rashes, severe muscle pain, sensitivity to light, and sometimes delirium.*

By now virtually all of the emigrants had given up on Meek, who for the most part was still riding a day or so ahead of the main party with his wife and a small

---

* Roughly a hundred years after it afflicted the Meek Party, typhus would kill thousands in Nazi concentration camps.

group of friends, perhaps as much to avoid a lynching as to find a way out of the nightmare he had led the others into. On occasions when he had to be in camp, his few remaining friends concealed him in their wagons.

At dawn on September 17, the twenty-fourth day out, they finally struck a reliable source of water at the south fork of the Crooked River, a tributary of the Deschutes, which they knew would lead them toward the Columbia. But they were far from out of danger. Large numbers of them were now too weak to walk or even to ride on horseback, so they had to be loaded into wagons pulled by fewer and fewer oxen. And they were still many miles south of The Dalles and the Columbia River.

Traveling out ahead of the rest of the party, Meek and his wife finally staggered into a Methodist mission at The Dalles on September 29 and informed the startled settlers there that hundreds of people were in dire distress in the interior. He purchased supplies for a relief expedition but declined to make the return trip to deliver them himself, presumably fearing for his life at the hands of the emigrants. Fortunately, someone else stepped forward.

Moses Harris, one of a very few African-American guides and mountain men, was also one of the best.

Called "Black Harris" or sometimes the "Black Squire" by his fellow mountaineers, he had traveled throughout the West since 1823, when he first crossed the Mississippi, probably as a freed slave. In 1844, Harris had guided a train of five hundred souls along the Oregon Trail all the way to Fort Vancouver. He had helped build Fort Laramie. He was thought by his peers to be unsurpassed in winter travel and survival skills. And now he was the only one at The Dalles who was both capable of, and willing to attempt, a rescue of Meek's lost party.

Harris and a small group of rescuers, many of them Indians from the mission, rushed south with provisions and, just as important, with equipment for helping the emigrants surmount one last challenge. To reach The Dalles, they had to cross the Deschutes River deep in its canyon, where it ran swift and cold and deadly between sheer basalt cliffs.

But Moses Harris had anticipated that and brought block and tackle, pulleys, ropes, and axes with him. He and the emigrants, with considerable help from local Indians, who were familiar with the ways of the Deschutes and who fished from elaborate scaffolding that they suspended over the river, set about building a suspension bridge for those who could still walk. Then they began caulking the beds of wagons, loading the sickest of the emigrants into them, and towing them across the water. It took two weeks to get everyone across.

Even after the last of them arrived at The Dalles, the emigrants who had taken the shortcut continued to die. A few were so hungry that they bolted down half-cooked food and became ill. Some were simply too weakened from the ordeal to have any chance of recovering. In the end more than fifty of them died.

As they finished the last of their preparations in St. Joe in late May, Sarah and Jay and Sarah's family learned that another party of Oregon-bound emigrants, the last of the season, was assembling on the far side of the Missouri just a bit to the north of town. Though he and his family were bent on California, not Oregon, Franklin Graves knew that it would be best to have company crossing the plains, so when they were fully provisioned, they drove four or five miles north of St. Joe to a crossing called Parrott's Landing and loaded their wagons onto one of the big, flat-bottomed scows that served as ferries.

When the boat pulled away from the Missouri shore, they left the United States behind them. Everything west of here was foreign and alien. There were no inns, no stores, no farms, no reliable means of resupply except for a couple of frontier forts hundreds of miles down the trail. But their three wagons were amply stocked, a small herd of beef cattle swam alongside the ferry, and fistfuls of silver coins were squirreled away under the cleats in their family wagon. Nobody, it must have seemed to them, could be better prepared for the journey ahead.

However, they had neglected one critical piece of advice. Of all the many tips, encouragements, admonitions, and suggestions that Lansford Hastings dispensed in *The Emigrants' Guide to California and Oregon,* the best of them had to do with timing one's departure. On this he was both honest and correct when he said that the emigrants must "enter on their journey on, or before, the first day of May; after which time they must never start, if it can possibly be avoided." On the consequences of not doing so, he was even more pointed: "Unless you pass over the mountains early in the fall, you are very liable to be detained by impassable mountains of snow until the next spring, or perhaps forever."

On the day that Sarah, Jay, and the rest of the Graves clan stepped aboard the ferry at Parrott's Landing, May Day was already more than three weeks in the past.

3

GRASS

When she disembarked from the ferry on the western side of the Missouri, Sarah found a movable village assembling itself in the woods. Perhaps 150 of her fellow emigrants were bustling around among tents and wagons, busily making their final preparations for the journey ahead. Men were shoeing horses and oxen, packing and repacking supplies, sharpening knives and axes, cleaning guns, gathering in small groups to study maps and consult guidebooks, talking politics, swearing, starting to get to know one another, sizing one another up. Women, too, were beginning to make acquaintances, edging up to one another and introducing themselves, gathering around smoky campfires to share folk remedies and recipes and gossip, trading small items and bits of advice. Some of them, like many of the men, smoked tobacco in long-stemmed white clay pipes as they talked.

Older children helped their parents with their various chores; younger children played in wagons or down along the river, skipping stones, exploring, looking for treasure, watching out for Indians—dangerous ones now that they had crossed the river. On the fringes of the camp, adolescent boys and girls eyed one another from afar and tended to livestock, talking to horses as they led them about, shouting at mules and oxen, snapping willow switches at their hindquarters to make them move. Here and there someone sat in the shade of a tree reading a Bible or a book of verse or writing a first excited and optimistic letter from Indian Territory to someone back home. A few plucked at banjos or sawed softly at fiddles, trying out tunes. From time to time, people stopped what they were doing for a moment to swat at the mosquitoes and horseflies that pestered them. Dogs ran to and fro, yapping and making acquaintances of their own, bounding down to the river and throwing themselves into the water for the sheer joy of it, then racing back ashore and shaking off the water, then bounding back into the river again. The smells of wood smoke, of frying bacon, of coffee, and of baking pies melded together and drifted among the wagons.

Franklin Graves led his family into this temporary village by the river and found a spot to park their wagons and pasture their cattle. Except for themselves, almost all of these families were bound for Oregon, or thought they were. That meant that at some point along the trail, at least by the time they reached Fort Hall, where the Oregon and California trails diverged, they would need to find new traveling companions. But that was months in the future, so for now they climbed down from their wagons and began to mingle, starting to forge their own friendships.

Among these friendships, those that would prove the most important, in unexpected ways far down the line, involved several families headed up by particularly large men—men who were outsize not only physically but, as it would turn out, also in the quality of their character and the quantity of their courage.

One of them was Colonel Matthew Dill Ritchie. Born in Pennsylvania in 1805, Ritchie, like Franklin Graves, had made a series of westward moves with his wife, Caroline, and an expanding brood of children. Like Franklin Graves, Ritchie had fought in the Black Hawk Indian War more than a decade earlier, and it was in that bloody event that he had acquired the title of colonel. Among Ritchie's children were two teenagers with whom the older Graves children could interact and find much in common as they traveled across the plains—fourteen-year-old Harriet and nineteen-year-old William Dill. Another of Ritchie's offspring, Mary Jane, was married to the largest of the large men in the camp—a twenty-eight-year-old giant named John Schull Stark. From an old Kentucky family, Stark was a distant relation of Daniel Boone. Powerfully built and weighing 220 pounds, he was said to have the strength of two men.

Yet another set of friendships that the Graves family began to develop in that first encampment by the Missouri revolved around Reason Penelope Tucker and his sons. Like Franklin Graves, M. D. Ritchie, and John Stark, Reason Tucker was a large-framed man. Gentle and soft-spoken, he wore his beard, as was popular at the time, in a fringe around his clean-shaven face. A Virginian born of Scottish parents, he had been married and widowed once already, and he now found himself a bachelor again as his second wife refused to follow him west. Traveling with him, though, were his three oldest boys—John Wesley, Stephen, and George Washington.

All three of these families—the Tuckers, the Ritchies, and the Starks—and most of the others that gathered by the Missouri that third week in May had much in common with the Graves family and with one another. Whether their ancestors first stepped ashore on the rich tidelands of Virginia or the stony shores of New England, they were almost all the children and grandchildren of men who had fought in the American Revolution. Many of them carried in their hands weapons that had been used in that conflict, and they carried deep in their hearts an absolute devotion to the idea that their liberty was the most valuable thing they owned. They commonly and solemnly referred to the fourth day of July as "the Glorious Fourth," without the slightest hint of irony or embarrassment. They named their sons Jefferson, Franklin, Washington, Lafayette, or Adams, lest those sons forget where they came from and how they had gained their unique and sacred freedoms. For the most part, they despised what they called "the trammels of civilization" and preferred to stay close to the frontier, even as it moved relentlessly westward. They believed deeply that they were destined to spread the light of liberty across the continent—to create, in fact, as Thomas Paine had put it, "the birth-day of a new world" in the West. They tended to be forthright, plain-speaking, earnest, friendly, and trustworthy. They took a man at his word, unless they had good reason not to. And above all they were fiercely self-reliant, unflinchingly independent. In the trying weeks and savage months ahead, though, they would find that one man's freedoms could become another man's fetters.

. . .

They broke camp and moved out on May 23, climbing up out of the Missouri bottomlands, following the course of a stream called Clear Creek. When they reached the top of the bluff, they got their first full view of what lay to the west. The blue-green prairie grass was knee-high now, still pushing up through the taller, dry, dead grass of the previous year. Gently rolling hills extended to the horizon, and the swell of those hills, along with the grass billowing in the wind, created the overwhelming impression—shared by nearly all who saw it—that they were about to set forth on a great, windswept sea.

In the first fifty miles, they had to cross a series of muddy creeks, some of them running in gullies etched as deep as twenty feet into the surrounding prairie. At many of them, they had to fill the stream with brush and then pull the wagons across with ropes. At Mission Creek they came across the last sign of American civilization that they would see until they reached Fort Kearney in present-day Nebraska—a thirty-seven-room, three-story brick Presbyterian Indian mission just being constructed that spring to replace an older log structure. West of the mission, they moved out into flatter, wide-open country.

For the first time, they started to see what many of the men in the party, and many of the women as well, had been looking forward to since leaving home, an astonishing quantity of wild game—turkeys, prairie hens, wild geese, elk, deer, and occasionally pronghorns—or "antelope," as they called them. They had not yet come to the buffalo herds, which everyone eagerly anticipated, but this was the beginning of an extended hunting excursion that many of them had dreamed of all through the preceding winter. There were no vegetarians among them. They were serious meat eaters all, and as they made camp each evening, they could now look forward to feasting on roast fowl, grilled steaks, rich stews, and pot roasts.*

On the fourth night out, Sarah's seventeen-year-old brother, Billy, took his turn standing guard on the perimeter of the camp, along with several other young men. It was a dark night, with only a thin crescent moon hanging low in the western sky. At about nine, Billy noticed that a grass fire had erupted in an arid patch of prairie about a half a mile north of the camp. The flames, pushed along by a dry west wind, streaked

---

* The following winter, one of the 1846 emigrants, Edwin Bryant, calculated and noted with some astonishment that his companions in John Frémont's "California Battalion" could happily eat an average of ten pounds of beef per day. And then ask for more.

by well to the north of the camp and seemed to pose no threat to it or to the livestock. Within a few minutes, though, he was surprised to see a young man running toward him full tilt. Billy called out, "Who comes there?"

"Friend!"

Billy recognized the man as one of his fellow guards but decided he'd best follow protocol.

"Friend, advance and give the countersign!"

"Don't talk so loud. Hain't you seen them?" the young man croaked.

"Seen what?"

"Why, the Indians. They are setting the prairie on fire and are going to surround us and kill us and take our stock."

"Where . . .?"

"Why, there, running along by the fire. There are hundreds of them."

The young man pointed at the flames. Seriously scared now, trembling in fact, as he later admitted, Billy turned and looked at the fire again. He was ready to bolt for the camp, but then he looked one more time.

In silhouette against the flicker of the flames, he saw the form of hundreds of what the emigrants called "resin weeds" bobbing and wavering in the wind.*

---

* Probably a species of *Grindelia*.

With the flames streaking past them, the tall weeds did appear to be human forms.

Billy laughed and pointed out the error, and the young man sheepishly extracted from him a promise not to tell the others about his mistake. But even as they chuckled about it, they heard a commotion in the camp. Dozens of men were rushing about, clutching guns, shouting, giving orders. Two other guards on the far side of the cattle had also seen the phantom Indians and rushed into camp to sound the alarm. For days thereafter much of each evening's entertainment centered on tormenting those other two guards about their hasty retreat in the face of marauding resin weeds. Billy and his companion sat smugly by and watched.

Anxiety about Indian attacks pervaded all the emigrant parties that headed west that spring. To a large extent, the fear was exaggerated and misplaced. Most of the Plains Indians that Sarah and her family would encounter along the way were not predisposed to attack the emigrants. They had little reason to hazard their own lives in order to take those of the emigrants, though that would change in the years ahead with the wholesale destruction of the buffalo herds and the utter destitution that the tribes began to experience as a result. For now, while they were often intensely curious about the various exotic foods and gadgets the white strangers carried with them, and sometimes coveted them, the Indians generally tried to obtain them through barter rather than the use of force. Some of the young men among them, however, were quite willing to make off with the emigrants' fat beef cattle and horses if given sufficient opportunity.

Most of the emigrants had developed very hard attitudes and deep prejudices well before they first encountered what they regarded as the "wild Indians" of the plains. For Sarah's family, as for many of those traveling with them, those hard feelings went back to the bloody sequence of events that they had experienced firsthand shortly after arriving in Illinois—the Black Hawk War.

In April of 1832, about a thousand Fox, Sauk, and Kickapoo Indians—men, women, and children—had crossed the Mississippi and entered Illinois bent on returning to their ancestral lands, territory that they regarded as sacred but had lost in a disputed treaty in 1804. A sixty-five-year-old Sauk warrior named Ma-Ka-Tai-Me-She-Kia-Kiak, or Black Hawk, led them. Black Hawk had hoped to avoid conflict, but his entry into Illinois set off widespread panic among the white settlers. Within days a makeshift assemblage of U.S. Army troops, Illinois volunteer militiamen, and Sioux and Menominee mercenaries was pursuing him. The militia, called out by the governor of Illinois, consisted of virtually all the healthy adult white males in Illinois. Throughout April and May of that year, small, detached bands of Native Americans, some of them only loosely allied with Black Hawk, fought a series of battles and skirmishes against the whites and their Native American allies.

On the afternoon of May 20, things got truly ugly. A group of seventy to eighty Pottawatomie warriors, apparently not attached to Black Hawk at all, attacked a white settlement of three families at a place called Indian Creek. The whites were quickly overwhelmed, and the results were horrific. A Native American witness later recounted what transpired.

> The women squeaked like geese when they were run through the body with spears or felt the sharp tomahawk entering their heads. All of the victims were carefully scalped; their bodies were mutilated and many of the children were chopped to pieces with axes; and the women tied up by the heels to the walls of the house; their clothes falling over their heads, . . . their naked persons exposed to the public gaze.

Fifteen settlers were killed. Fifteen-year-old Rachel Hall and her seventeen-year-old sister, Sylvia, crawled into a bed and tried to conceal themselves but were discovered and taken away as prisoners. That night they watched in horror as their mother's scalp, among others, was scraped and stretched on a willow hoop to cure.

That same day, in Lacon, Franklin Graves answered the governor's call and joined the militia. His neighbor John Strawn had the previous year been named a Colonel of Militia. Now Strawn, attired in a full regimental dress uniform replete with gold epaulets and a plumed helmet, stood on an open piece of ground, formed Franklin Graves and the other men of the neighborhood into a line, and addressed them with a flourish. "Ye sons of thunder! Our country is in danger, and the call is 'To arms.' Those willing to enroll yourselves among her defenders will step three paces forward." Franklin Graves stepped forward and enrolled as a noncommissioned officer, the outfit's drum major. For the next month, he and the other members of the newly constituted Fortieth Regiment of Mounted Volunteers drilled and marched up and down the Illinois River searching for hostile Indians but finding none.

As the men marched and searched, the women back home grew increasingly nervous. The first night after the men's departure, a group of women in nearby Richland Township came together at Nancy Dever's house to discuss how best to defend themselves if attacked. The women were not at all happy that when the men had strutted off to war, they had taken with them nearly every household gun in the township. The women decided that if they were to make it appear as if the Dever house had already been attacked and ransacked, any marauding Indians might pass it by. So they scattered an assortment of furniture and linens and other household goods haphazardly about in the yard. Then they took some food and bedding up into the cabin's loft and lay low.

Later that evening, at dusk, more neighborhood ladies arrived at the cabin. Seeing the shambles in the yard, they concluded that the Devers had been massacred and let out ear-piercing screams of lamentation. The women in the loft, hearing the screams, took them for Indian war cries and, believing they were about to be relieved of their scalps, unloosed a salvo of answering screams. Eventually one of the women in the cabin peered outside and realized the true situation. The ladies spent the rest of the evening sheepishly hauling the Devers' household effects back into the cabin.

Eleven days after the abduction of the Hall sisters, following some negotiations conducted via Winnebago Indian intermediaries, the girls were ransomed for a bit of money, ten horses, and some corn. Over the next several months, nearly one-third of the U.S. Army and nine thousand Illinois militiamen pursued Black Hawk and an ever-diminishing band of his supporters northward into Wisconsin. After a series of one-sided battles, Black Hawk tried to surrender on August 1, wading into the Mississippi and shouting his intentions to the *Warrior*, a steamboat that had been chartered by the army. No one on board could understand him, though, so they opened fire, killing two dozen of Black Hawk's men but missing him.

The next day the troops surrounded Black Hawk and the last four hundred of his people, and what is somewhat euphemistically called the Battle of Bad Axe began on the eastern side of the Mississippi. As Black Hawk's men fought to hold off the troops, the women tried to swim across the river, many of them with their children clinging to their backs. The men on the *Warrior* shot them one by one. When there was no one left to shoot in the water, the *Warrior* unleashed its cannons on the people clustered along the shore. Then the troops closed in. They shot nearly everyone still alive. They shot old men, women trying to surrender, children trying to flee. By the end of the day, there were hundreds of bodies scattered along the banks of the Mississippi. The troops scalped most of them. Then they cut long strips of skin from the backs of some, in order to make razor strops.

Black Hawk himself escaped, but he surrendered at Prairie du Chien on August 27. He was imprisoned for a time, but in 1833 he was returned to what remained of his people, settled now in Iowa, and he died there in 1838. However, no one who had experienced the violent events of 1832 on the Illinois frontier forgot about them, on either side.

Sarah and her family pressed on westward under mostly gray skies, following a snakelike road that wound its way along the crests of the low hills separating the drainages of the Missouri River and the Kansas River. The country was increasingly open now. Thunderstorms began to pop up, drenching the prairie, swelling the rivers, and hastening the growth of the tall, rippling, blue-green grass.

A hundred miles to the west, and slightly to the south of them, the Donners, the Reeds, and the rest of the Russell Party were stalled at a pretty spot that one of their number, Edwin Bryant, had just named "Alcove Springs" on the eastern shore of the Big Blue River. The thunderstorms had raised the river far too high to ford. For days the travelers had sat in the rain, doing laundry, watching flotsam race by at fifteen miles an hour, and waiting for the river to fall. Finally they had begun to construct a ponderous and awkward log ferry.

On May 29, James Reed's seventy-year-old mother-in-law, Sarah Keyes, died, necessitating a funeral and a delay in working on the ferry. Keyes's granddaughter Virginia, who had turned thirteen the day before, wrote her cousin back home about the funeral.

We buried her verry decent. We made a nete coffin and buried her under a tree we had a head stone and had her name cut on it and the date and yere verry nice, and at the head of the grave was a tree we cut some letters on it the young men soded it all ofer and put Flores on it We miss her very much every time we come into the Wagon we look at the bed for her.

On May 30 they completed work on the ferry. On the thirty-first, after working from dawn until 10:00 P.M., struggling with the heavy ferry in the swift current and a drenching rain, they finally got the last of the wagons across. But the work was brutal, wet, and cold, and that night a fight broke out in their camp. Tensions had arisen between some of the California-bound emigrants and those bound for Oregon. Knives were pulled out but then put away. Nerves were beginning to fray, and they were not yet out of modern-day Kansas.

. . .

By June 7 the weather across the central prairies had finally begun to warm up and dry out, the thunderstorms growing less frequent and the turgid rivers finally falling. Still well to the east of the Russell Party, Sarah and her family—along with the Tuckers, the Ritchies, the Starks, and the other emigrants out of St. Joe—were making steady if slow progress, traveling at the speed of plodding oxen, working their way across Kansas just south of the present Nebraska line. Then, turning northwest, they descended into the Platte River Valley and intercepted the Great Platte River Road, the 1840s version of an interstate freeway, near Fort Kearny in Nebraska.

The daily routine had become monotonous, and the novelty of camping out had worn thin, but, for the most part, Sarah and her siblings were still in high spirits. By day they rode or walked through waist-deep, verdant grass. Meadowlarks, clinging to the tall blue stems of grass or to shrubs, chortled and sang, bobbling on their perches in the ceaseless wind. Grasshoppers rattled away from under their feet, flashing flame red or canary yellow underwings. Orange and white and yellow butterflies drifted above the grass. Metallic green and blue dragonflies darted and dodged over the white canopies of the wagons.

On warm evenings, after the livestock were all taken care of and the aftermath of dinner had been cleared up and the guards had been posted, the young people gravitated toward one another, congregating near one campfire or another. Sometimes they just sat in the dark and listened to the night sounds—crickets singing out in the grass, frogs shrilling along a creek, prairie wolves howling and coyotes yapping off in the distance. Usually, though, they talked about the day's events, about the novelty of the scenery or the wild animals they had seen that day, or about what California was going to be like when they finally got there. They were, by and large, still getting to know one another, and all of them were conscious of a fundamental fact that was at once both sobering and thrilling for those among them who were not yet married: They had entered into a very small world, a world where a substantial portion of the young men and women that they would encounter in the future, even after they reached Oregon or California, were sitting here at the campfire with them. Many of them were likely to marry one another, either by choice or by necessity, and they knew it.

So they would chat nervously, watching out of the corners of their eyes to see who was sitting close to whom that night. Then Jay Fosdick, or another of the young men, would get out his fiddle, and somebody else would get out a guitar or a banjo. They would begin to reel off some tunes—"The Arkansas Traveler," "Money Musk," and the popular "Virginia Reel." The Graveses' teamster, John Snyder, big and handsome, would drop the rear gate on one of the wagons, climb up onto it, and dance a wild jig while the girls laughed at his antics and the boys clapped time. Two by two, boys and girls would get up and pull off their boots and begin to dance, circling and wheeling around the fire, their bare feet kicking up dust, their warm hands holding other warm hands, or lying lightly on broad shoulders, or pressing against the delicate hollow at the small of a slender back.

Tune after tune would fly off the fiddles—fast, hot tunes that swirled through the night air and made you want to get up and stomp your feet. Older folks clutching tin mugs full of coffee would drift toward the fire to watch and then set down the coffee and join in, flinging out their arms and legs like drunken chickens as the screeching of the fiddle and the twang of the banjo stole into their souls and rattled their bones. Somebody would start singing, and a chorus would join in, all of them sharing in songs that had come down to them from their grandparents in Scotland or Ireland or the hills of Virginia or New England—boisterous sea chanteys, sweet love songs, uplifting spirituals, funny songs about foolish lovers or clueless dandies, and many, many sad songs about the folks left back home.

When they had worn themselves out with singing and dancing, the older people would drift away toward their tents or wagons. But the young would linger, sitting around the fire again, staring into it, enjoying the feel of its heat on their faces as the prairie night began to grow chilly, throwing a few more sticks onto it, listening to them pop and hiss. They would talk again, more quietly now that people were trying to sleep nearby, the boys joking and boasting, the girls giggling at them or ignoring them, depending on the signals they wanted to send. Hearts began to grow fond. Some later said that Mary Ann Graves and John Snyder had begun to take particular notice of each other.

Romance was blooming on the road out ahead of them as well. In light of the tension between the Oregon-bound emigrants and the California-bound, the Russell Party had split into two groups on June 2, the Oregonians traveling a few miles out ahead of the Californians under their own captain, Rice Dunbar. On June 12 a member of the California contingent, thirty-five-year-old Charles T. Stanton, a bachelor from Chicago, wrote about the effect of the split on the younger people in a letter to his brother.

In their party there were many young ladies—in ours mostly young men. Friendships and attachments had been formed which were hard to break; for ever since, our company is nearly deserted by the young men riding out, riding on horseback, pretending to hunt, but instead of pursuing the bounding deer or fleet antelope, they are generally found among the fair Oregon girls! Thus they go, every day, making love by the road-side, in the midst of the wildest and most beautiful scenery. . . .

And indeed, within the next few weeks, three of the Oregon girls were married and two more engaged.

In the same letter, Stanton—who though he stood only five feet five would in the fullness of time turn out to be among the biggest of the big men of the 1846 emigration—also wrote about something more ominous that had befallen the Oregon contingent a few days earlier.

This little party, one day before they reached the Platte were surprised by a band of 20 or 30 Pawnees, drawn up in battle array, coming down full sweep to attack them; but they were no sooner seen than the men formed in order of battle to meet them. The cunning Pawnees, seeing this little band

*drawn out, and fearing the deadly rifle, immedi-
ately turned their war party into a visit—shaking
hands, hugging men, and attempting to embrace
the women.*

The Pawnees had a bad reputation among many of
the emigrants of 1846. Rumors about them abounded
on the trail, and J. M. Shively, author of one of the most
popular of the trail guides, had singled them out as a
particular threat.

Not everyone in the loose assemblage of wagon trains
spread out between the Big Blue and the forks of the
Platte River in the middle of June was worried about
the Pawnees, though. On June 16, Tamzene Donner
wrote to a friend back in Springfield that aside from
all the cooking she had to do, she was thoroughly enjoying herself—reading the collection of books she had
brought along, taking in the novel scenery, and particularly botanizing. The prairie was in full bloom now,
and she was enthusiastically sketching and preserving
dozens of species of wildflowers—lupine, "ear drop,"
larkspur, poppy mallow, wild hollyhock, and many
others she could not name. As for the Indians, she reported that some local chiefs had taken breakfast with
the Donners at their tent that morning and that "we
have no fear of Indians. Our cattle graze quietly around

our encampment unmolested. Two or three men will go hunting twenty miles from camp; and last night two of our men lay out in the wilderness rather than ride their horses after a hard chase." Then she summed up her state of mind in words that would eventually grow monumentally ironic and chilling: "Indeed if I do not experience something far worse than I have yet done, I shall say the trouble is all in getting started."

PART TWO

The Barren Earth

And nothing can we call our own but death
And that small model of the barren earth
Which serves as paste and cover for our bones.

—WILLIAM SHAKESPEARE,
King Richard II

4

DUST

As Sarah and her family made their way up the Platte River in June, James K. Polk's plan for California was unfolding nicely out west.

Aside from John Sutter, about whom Vallejo fretted a good deal, Don Mariano Guadalupe Vallejo was the undisputed ruler of the Mexican province of Alta California in 1846. The official military governor of the province and owner of more than 175,000 acres of the finest agricultural land in the world, he was estimated to have an income of ninety-six thousand dollars a year, an enormous figure in the mid-nineteenth century. His empire was centered in a fortified compound and a two-story adobe house he called Casa Grande in the sleepy town of Sonoma north of San Francisco Bay. There he maintained a handful of Mexican troops, equipped with nine brass cannons, 250 muskets, and a hundred pounds of gunpowder, everything that the Mexican government considered necessary for the defense of the province.

Born in 1807 in Monterey, California, Vallejo was thoroughly Mexican, but in recent years he had watched in frustration as a series of revolutions had racked Mexico and the resulting succession of governments had left California to drift aimlessly. Increasingly, Vallejo, like many Californians of his generation, had begun to wonder whether the province might be better served if it were entirely free from Mexican governmental control. He was on good terms with many of the American settlers who had arrived in his neighborhood in the previous few years, and the idea that California might become an independent American protectorate had begun to intrigue him.

So it was with some considerable astonishment, but also some sense of inevitability, and perhaps even some relief, that Vallejo awoke just after dawn on June 14 to the sound of men banging pistol butts on the main door of Casa Grande. Vallejo and his wife, Doña Francisca, peered out of an upstairs window. In the large, dusty courtyard below, a motley assemblage of men with rifles and pistols were shaking their fists and shouting curses and threats in English. Vallejo hurriedly pulled on his dress uniform and went downstairs. Doña Francisca urged him to flee out a back door, but Vallejo instead went to the front door and flung it open.

The men who stormed into the elegant great room of Casa Grande were dressed in ragged buckskins, torn blue breeches, coyote-skin hats, and greasy red bandannas. A few were bare-chested. Some wore muddy boots; others were barefoot. They smelled of horse sweat. Tomahawks and pistols dangled from their belts, notched to show how many men they had killed. Their apparent leader, one Ezekiel Merritt, wore a long beard stained dark brown with tobacco juice. They were freelance revolutionaries, loosely and unofficially associated with a U.S. Army officer, Lieutenant Colonel John C. Frémont, who was now encamped just outside of Sutter's Fort in the Sacramento Valley.

Vallejo addressed them formally: "To what happy circumstance shall I attribute the visit of so many exalted personages?" The exalted personages chewed for a moment on their tobacco and then told Vallejo, somewhat less formally, that they had come to arrest him, to secure the surrender of Sonoma, and to declare California an independent republic. Vallejo nodded and ordered someone to bring out his best wine and his best brandy. Then he invited the surprised revolutionaries to sit down for a drink.

Together they began to draft a document formalizing Vallejo's surrender and California's declaration of independence. By 8:30 A.M., the wine and brandy had mellowed everyone out considerably. The proceedings inside Casa Grande took on the atmosphere of a party. The revolutionaries were hungry, so Vallejo had a young steer butchered, and breakfast preparations were begun. By 11:00 A.M., the document was complete, and California was a republic—at least in the eyes of the revolutionaries. Vallejo went to his room and fetched a ceremonial sword and handed it over to the men in buckskin, but nobody could figure out what to do with it, so Vallejo returned it to his room. Finally Vallejo was ushered out of his house to be taken under guard to meet Frémont at Sutter's Fort.

Out in the courtyard, William Todd, a nephew of Mrs. Abraham Lincoln, sat down on the ground by the flagpole with several of his fellow revolutionaries, a piece of unbleached cotton, some scraps of red flannel, and some homemade ink made of linseed oil and ferric oxide. If this were to be a republic, they figured it ought to have a flag. They emblazoned the cotton with the words "California Republic." Above that they drew a star and what they intended to be the figure of a grizzly bear. Then they ran the flag up the pole. The Mexican Californians who had gathered around, suddenly foreigners in their own land, looked up, pondered it silently, and wondered why the Americans had chosen a pig as the symbol of their ascension to power.

The American seizure of California would not in the long run turn out to be a laughing matter, though, nor a bloodless one. A few days after Vallejo's capitulation at Sonoma, two young American men named Cowie and Fowler were apprehended by a small Mexican force near Bodega Bay. After being kept prisoner for a day or two, Cowie and Fowler were tied to a tree and stoned. When one of them suffered a broken jaw, a cord was tied to the jaw and it was yanked out of his face. Then, slowly and deliberately, their tormentors began to slice off bits and pieces of their flesh—fingers and toes and other appendages—and stuff them down the men's throats. Finally the two men were disemboweled and left to die.

On June 17, the day after Tamzene Donner had taken breakfast with some local Pawnee chiefs and written her buoyantly optimistic letter home, Sarah and her party arrived at the same approximate spot on the south bank of the Platte River. That night a party of Pawnees stole up to the edge of their encampment and ran off a large portion of the emigrants' 150 head of cattle.

The loss of so much livestock could prove catastrophic later in the trip, so in the morning small groups of men fanned out across the surrounding countryside to round up as many of the cattle as they could find. As two of them—an Iowan named Edward Trimble and a Ohioan named Harrison—were driving five head of the cattle back to camp, a dozen or more Pawnees rose out of the tall prairie grass, demanding the men's horses. Trimble refused, and the Pawnees promptly loosed a volley of arrows and rifle shot at him. Trimble tumbled from his saddle, dead. The Pawnees took Harrison prisoner and were in the process of stripping him naked when another search team arrived on the scene and charged the Pawnees, chasing them off.

Trimble left a pregnant wife, Abarilla, and four small children. Faced with the prospect of continuing across the plains pregnant, with young children in tow and without a husband, she declared that she wanted to go home to Iowa. The only eastbound travelers in the area, however, were all male. A woman could not even contemplate giving birth without the assistance of other women, so she reluctantly continued westward.

Trimble's death, and the plight of his widow, must have abruptly changed the prevailing mood of Sarah's party, particularly that of the women. Every wife among them was suddenly brought face-to-face with a hard reality that they generally tried not to think about but constantly feared nevertheless. To be widowed out here on the plains or in the mountains ahead was—to a large extent—to be rendered instantly dependent on the goodwill of the men around you. As good and trustworthy as those men might be—and most of them were both—few of them had extra time or energy to drive your cattle, to hunt game for you, to repair your wagons, to chop your firewood, or to attend to the various other heavy chores that you might require having done. Most of the men either had families of their own who had to be a first priority or were hired hands, expected to tend to the needs of their employers. Bereaved widows were sometimes incorporated into other families to some extent, and the other women in the party almost always gave them extra help. In the end, though, the amount of energy and the number of resources that any family could muster were finite, and there was a practical limit to which either could be shared. Farther down the trail, everyone knew, a shortage of either could be a matter of life or death.

Many emigrant women who found themselves in this precarious situation adapted themselves to it heroically, doing all the "masculine" chores themselves. But it was a heavy, sometimes crushing, burden to add to the traditional duties they already performed.

It was all the worse for those women who were mothers as well as wives, as most of them were. The number of ways in which their children might come to harm along the trail was staggering, and women who had to drive a team or repair a wagon were unable to devote much time to watching out for them.

Children fell under wagon wheels and were crushed to death or crippled for life. They wandered off into the tall grass and were never seen again. Occasionally they were abducted by Native Americans. Much more frequently they drowned when swept away by rivers their families were trying to ford. Drowning incidents were so common, in fact, that some mothers wrote their children's names in indelible ink on labels and sewed the labels into their children's clothes. It didn't prevent them from drowning, but it sometimes allowed a grieving mother to identify a body that had been in the water too long. Children were bitten by rattlesnakes, struck by lightning, trampled by unruly oxen or horses, pummeled by hailstones as large as turkey eggs, and shot by the nearly daily accidental discharges of the guns that their fathers carried. They died of measles, diphtheria, whooping cough, influenza, tuberculosis, typhoid fever, malaria, infected cuts, food poisoning, mumps, and smallpox. Perhaps the only break that mothers on the Platte River Road had that summer was that it wasn't yet 1849, when Asiatic cholera would kill thousands along this same stretch of trail, the graves in some places averaging one every two hundred feet.

The greatest fear that widowed mothers shared, though, was that they themselves would die and leave their children orphaned on the trail. Usually other mothers took such orphans in. One woman, Margaret Inman, walked five hundred miles with an orphaned newborn in her arms, searching out nursing mothers who would offer the infant their breasts each night and morning. But as the emigrants got farther west and every aspect of life on the trail got harder, attitudes also hardened. When both got hard enough, nobody could count on compassion. It didn't happen often, but more than once during the emigration years of the late 1840s and early 1850s, women who did not own or could not drive wagons were simply left behind as their companions moved on over the horizon.

It happened to a young woman named Polly Owen White. When her husband died, the family for whom he had been working refused to take her any farther without some form of payment. When she couldn't produce it, they left her and her baby girl sitting by the side of the road. Eventually they were picked up by another family and made it to Oregon, where Polly remarried and went on to bear, improbably, four sets of twins. But everyone knew that under extreme circumstances individuals would look out first for themselves and their family members and that abandonment of the old, the ill, the lame, the helpless, or the dependent was always a possibility.

As the Graves family and their companions moved up the Platte River Valley and deeper into Pawnee country in late June, they increased their vigilance, posting guards all night, every night now. Following one day's hard work and preceding another as they did, the stints of guard duty were onerous, so they were rotated among all the able-bodied men. Women were not expected to serve guard duty, but when Jay Fosdick's turns came around, Sarah—with no children to attend to in her wagon—often accompanied him out into the dark.

It offered her a rare chance to be alone and unobserved with her husband, to do the things that young brides most want to do—to sit side by side with him and talk quietly about their future, to listen as he serenaded her softly with his fiddle, to tease him, to hold his hand, to lean against him and sit in the warmth of his arms. The moon had waned to a thin crescent by the third week in June, and the black velvet of the prairie sky was spangled with shimmering drifts of silver stars. When they were sure that it was safe, Sarah and Jay could sink into the prairie grass, lie on their backs, listen to the crickets, stare into the immensity of the heavens, count falling stars as they streaked across the skies, ponder what such things meant. They could pull each other close, brush warm lips, and make love quietly, lost in each other and in the vastness of the dark prairie night.

For the most part, sex in the 1840s was like sex anytime, but it involved complexities and carried consequences in 1846 that can be hard for us in the twenty-first century to fully appreciate. Sarah had come of age during one of the successive waves of evangelical Christianity that swept over the United States in the nineteenth century. Carrying strains of Puritanism, but also drawing on much older fundamentals of Western theology, the evangelical movement of the 1830s taught that sexuality in any form was essentially degrading. Virtuous people were those whose conduct was most removed from any association with the rest of the animal kingdom and the various animal appetites that could be readily seen in any barnyard—the sexual appetite in particular.

That was the official line at any rate. It was the implicit and explicit message that Sarah and her sisters and young women like them had likely heard many times at camp meetings and in country churches back home in Illinois. No doubt it profoundly affected the way they thought about sex and about themselves as women. But, of course, nature does assert itself, and sooner or later many country girls and the country boys with whom they had grown up did what country girls and boys have always done. They nestled under blankets on back porches, they crept into haylofts, or they snuck off into the woods.

The practical problem, of course, was what to do about unintended consequences. For Sarah, just married, the risk of pregnancy might not have been an issue; in fact, a first child was likely a reward that she and Jay looked forward to. But to be pregnant on the emigrant trail was to take on a very heavy burden, both literally and figuratively. It could easily mean the difference between surviving the trip and not, so conceiving a child was something that Sarah might well have wanted to postpone until she and Jay reached California. And for other young women on the trail that summer, single women who found themselves in the prairie grass or in the backs of wagons with young men that they fancied, the threat of pregnancy was an ever-present terror.

Married or not, there were options for a young woman who wanted to protect herself or who found

herself compromised, but they were closely held secrets. To learn them she needed to be pulled into a
circle of older women.

In the deeply paternalistic world in which they lived,
and the particularly masculine world of the wagon train,
the emigrant women of the 1840s had one powerful
thing going for them—a covert culture perpetuated by
the whispered counsel and shared confidences of other
women. On the road to California or Oregon, even
more than back in their frontier villages or on their iso-
lated farmsteads, women came together every day over
their common tasks. Doing laundry on a riverbank,
tending to infants under shady trees, keeping an eye
on older children playing among the wagons, cooking
on a marathon scale around campfires, they gathered
in small groups to work. And while they worked, they
exchanged more than recipes. They quietly exchanged
bits of information that few of their men knew, or would
ever know.

They talked of pessaries and syringes and sponges.
They exchanged formulas for herbal concoctions that
might, or might not, alleviate female problems. They
drew on knowledge that their mothers and grand-
mothers had shared with them when they were themselves first pulled into similar circles of women in
their villages back home. Much of the knowledge was

wrong, some of it dangerously so, but it was what they had to work with.

The options for trying to avoid pregnancy in the first place were particularly limited. Condoms in the 1840s were still handmade of sheep gut. As a consequence they were expensive and had to be carefully washed, dried, and cared for between uses. Few men had the patience for such things, so country women seldom saw them. Married women like Sarah sometimes tried to time their sexual encounters to avoid their most fertile periods, but their men were frequently unaccommodating in that regard. And even when their men cooperated, the women's own understanding of the biology behind their cycles was sketchy at best and fundamentally flawed at worst, often causing them to have sex during what were in fact their most fertile times.

So they tried other things. They cut sponges into small pieces, attached ribbons to them, moistened them with water or with mixtures they thought to be spermicidal, and inserted them prior to sex. Afterward they douched with fluids of all sorts, anything they thought might kill or immobilize the "amicules" that brought on pregnancy. Many of the douching mixtures were made from common ingredients—vinegar and baking soda in particular. Others were more exotic herbal concoctions or chemical solutions made with sulfate of zinc, boric acid, bichloride of mercury, or carbolic acid. To the extent that they changed the pH of the vagina, some of these may have been minimally effective, but as often as not, the application of the douche simply had the effect of forcing the sperm deeper into the vagina, closer to the cervix and to the ovum.

When it came to terminating unwanted pregnancies, though, women of the 1840s had one practical advantage over women today, and it was, paradoxically, a product of their ignorance. Little was understood of the biology of conception and gestation. Pregnancy was not generally thought to commence until the "quickening" began—the time at which a woman began to feel movements of the fetus within. As a practical matter, this meant that a woman often did not consider herself assuredly pregnant until the second trimester. The changes that attend early pregnancy—swelling of the breasts, tenderness, nausea, and particularly the cessation of menstruation—were generally attributed simply to a "blockage" of the woman's reproductive organs. Certainly, many women in this situation knew that they were in fact pregnant, but the "blockage" theory allowed them a certain amount of leeway to act aggressively to undo what had been done without wrestling with the moral implications. To restore menstruation they drank herbal teas, soaked their feet in hot water infused with herbs, or mixed herbs with mustard plasters and applied them to their breasts and bellies.

They often tried vigorous exercise—particularly anything likely to strain and jostle the abdomen, such as horseback riding. If these remedies failed to produce results, they resorted to ingesting strong purgatives and cathartics to promote severe cramps and vomiting. Some of these came in the form of the patent medicines that were just then beginning to become popular in the United States. More often they were homemade. Either way, many of them contained ingredients that were outright poisons—nightshade, oil of tansy, aloe, oil of savin, foxglove, mistletoe, and oil of cedar among others.

One of the most common and powerful of these agents was oil of hemlock, which in small doses was used to treat everything from sore throats to aching bones. In larger quantities, though dangerous, it was known to clear "blockages"—to induce miscarriages. It is one of the things that we know that Elizabeth Graves carried with her on the road to California. She might well have brought it along primarily to treat minor ailments, but with a bevy of teenage daughters in tow, she might also have thought of it as a wise precaution against potential trouble of a greater sort.

As Sarah and her family made their way up the Platte River toward Fort Laramie, the country gradually grew drier and less forgiving. Dust afflicted them now—dust so deep it sometimes came up to their ankles as they walked alongside their wagons, so fine that plumes of it rose behind their wagons and hung in the air, blinding everyone except whoever was lucky enough to be driving in the lead that day. When the wind blew, as it almost always did, the dust swirled around them in clouds so thick that they could not see the fronts of their own teams, let alone the wagons farther ahead. In the midday heat, it filled their nostrils; at night it settled on their food and made it taste like earth. It clung to their hair, infiltrated their clothes, stuck to their sweaty skin, became almost a second skin to them.

In far western Nebraska, trees and shrubs began to give way to sagebrush, and firewood soon became scarce and then all but nonexistent. Sarah and her mother and sisters had to build their cooking fires at night using dried buffalo dung for fuel. The shallow water along this section of the Platte was slow and sluggish and often stagnant, but they had no choice except to use it for drinking and for cooking. Sometimes they had to dig holes in the sand alongside the river to get reasonably clear water, but even then it often tasted bad and smelled foul. Small, low-growing cacti began to appear among the sagebrush, and walking with bare feet—as the travelers often preferred to do to save their shoes—became nearly impossible. Rattlesnakes also began to appear, slithering across the road or coiled up in the shade under sagebrush, something new for mothers to worry about.

Far to the west that same week, on June 25, the *Oregon Spectator* reported that an angry meeting of settlers had taken place at a hotel in Oregon City ten days before. The Oregonians had gotten wind of Lansford Hastings's plan to divert Oregon-bound emigrants to California, and they were furious about it. They had resolved to send express messengers to meet the emigrants, "in order to prevent their being deceived and led astray by the misrepresentations of L. W. Hastings. . . ." The messengers were to carry testimonials bearing witness to Oregon's fertility and benign climate, the comparative aridity and sterility of California, the ambition and duplicity of Lansford Hastings, and the hornet's nest of trouble that he seemed bent on stirring up with the Mexicans in California. The *Spectator* quoted from one of the testimonials of someone who had met Hastings in California.

When we left California for Oregon, Lansford W. Hastings started to meet the emigration from the states, to try to persuade them to go to California.

He told us publicly that he and Capt. Sutter intended to revolutionize the country, as soon as they could get sufficient emigrants into California to fight the Spaniards; this plan was laid between Capt. Sutter and L. W. Hastings before said Hastings published his book of lies. . . .

Hastings was in fact closing in on his rendezvous with the approaching emigrants. He and Jim Clyman had labored across the salt flats of Utah and made their way tortuously through the narrow canyons and brush-choked valleys of the Wasatch Mountains on mules and horses, arriving at Fort Bridger, a squalid trading post in southwestern Wyoming, on June 7, only to find it deserted. There they had separated. Clyman was by now thoroughly disillusioned with Hastings's so-called shortcut, but Hastings himself was still determined to intercept any Oregon-bound emigrants he could find and divert them to California via his new route, though he must surely have seen that it was nearly impassable for wagons.

On June 27, Clyman, traveling eastbound, arrived at Fort Laramie. That same day the Donners and the Reeds and those traveling with them—still about a week ahead of Sarah and her family—also arrived at the fort. Clyman was an old acquaintance of James

Reed's from the days of the Black Hawk War, when they had both served in the same company as Abraham Lincoln.

When the two men sat down together, Reed asked Clyman about the route ahead, and in particular about the cutoff he had read of in Hastings's book. Clyman was unambiguous. He told Reed to "take the regular wagon track and never leave it—it is barely possible to get through if you follow it—and it may be impossible if you don't." Reed protested that the established road was too far out of the way, so Clyman elaborated on the difficulty of the country he had just passed through—the confusing maze of canyons in the Wasatch Mountains, the nearly impenetrable brush, the devastating heat and aridity of the Great Salt Desert, and the snow and granite cliffs of the Sierras. But Reed was unconvinced. "There is a nigher route," he said, "and it is of no use to take so much of a roundabout course." The next day they parted, Reed and the Donners moving on west toward Laramie Peak and the South Pass through the Rocky Mountains, Clyman heading east, continuing to meet the last stragglers of the 1846 emigration.

By now Sarah and her family had entered an audacious landscape. They had crossed the bleak and tinder-dry hills between the south and north forks of the Platte, descended into a relatively lush glen called Ash Hollow, and begun working their way up the North Platte. Large bluffs and improbable rock formations sculpted by wind and ancient floods began to appear. Many of the rock formations were famous landmarks that Sarah and Jay had heard about along the trail. All of them seemed exotic and romantically beautiful to wide-eyed midwesterners like themselves, who had never in their lives seen anything taller than a fair-size hill. In the evenings, as the sun set and the sky flared orange and red and then deepened into violet and indigo, the rock formations all but glowed, bathed in shades of tangerine and burgundy.

They passed stately, monumental Courthouse Rock. Then, from miles away, long before they arrived at its base, they began to get glimpses of perhaps the most famous landmark on the trail, Chimney Rock, a thin stone spire arising from a conical base and soaring nearly five hundred feet above the surrounding desert. Finally they arrived at the largest of the landmarks, the fortresslike promontory called Scotts Bluff.

That evening they met Jim Clyman south of the bluff. There, in the gloaming, under the great peach-colored stone ramparts of the bluff, Clyman made it clear to Franklin Graves, as he had to many of the

California-bound emigrants he'd met on his way east, that he thought California a poor place to settle. Whatever he said, or did not say, about Hastings's cutoff that night was never recorded, but it's likely he told Graves what he'd told all the others—that it was a poor route.

By July 2, a hot, humid day following a night in which thunderstorms had rumbled and flashed above the rock formations just to the northeast of them, Sarah and her family were approaching Fort Laramie. The next day was her sister Lovina's twelfth birthday, and that afternoon they arrived at the fort, a stopping point that provided Sarah and her family with both a welcome relief and a considerable spectacle.

The fort itself was not particularly impressive—a square structure, perhaps 130 feet to a side. It stood on a low bluff in an oxbow of the Laramie River, just a few dozen feet back and uphill from the water. Its walls were made of adobe and rose fifteen feet tall, with blockhouses at the corners and a gate at the front entrance. The walls surrounded an inner plaza lined with storerooms, a blacksmith shop, and simple living quarters for the staff of the American Fur Company, who maintained the establishment. What those who were there that first week in July 1846 later remembered and wrote home about, though, wasn't the fort but the scene surrounding it.

A war party of more than two thousand Sioux had been moving toward the fort for days, on their way to do battle with their rivals the Crow. They had brought their families and their entire villages with them, dragging their possessions on sleds called travois, pulled by ponies and dogs. They were now encamped all around the fort, hundreds of their tepees set up in the groves of ash and cottonwood trees that bordered the Laramie River and offered shady relief from the bone-dry hills surrounding the fort. Smoke from their campfires mingled with that of dozens of fur trappers and mountain men who were also encamped here on their way into or out of the Rockies. The trappers, especially, were sometimes hard to distinguish from the Sioux. Mostly French Canadians or Yankees, some of them wore buckskins and their long hair braided, and many traveled in the company of their Native American wives and children.

In other groves and in the dusty open fields in front of the fort, scores of white-topped emigrant wagons were assembled in loose clusters. Everywhere men and women, both white and Native American, were bustling about—tending to children, trading horses for mules or tobacco for furs, doing laundry in the river, repairing wagons, sitting in the shade smoking clay pipes. The Sioux warriors—some of them bare-chested, some resplendent in suits of white-tanned buffalo skins decorated with colored beads, some wearing feathered headdresses and full war paint, some with scalps dangling from their waists—crowded into the small, dark, and relatively cool apartments of the fort itself, taking in the wonders of the commissary and trying to arrange favorable trades for rifles, knives, pots and pans, and provisions of all sorts.

The following day was the Glorious Fourth of July, in many ways the most important day of the year for the emigrants. At Fort Laramie and all along the trail, the day was given over to festivities. Bottles of whiskey that had been squirreled away in wagons were brought out and opened. Children gathered berries for the making of pies, tarts, and cakes. Men rode gleefully out on buffalo hunts or slaughtered beef cattle and prepared huge barbecue fires. They drank more whiskey. They hoisted the flag and fired salutes to the Stars and Stripes with the fort's cannons and with whatever other firearms they could get their hands on. They invited the Sioux into their camps and gave them small gifts. As evening came on, they built bonfires and drank whiskey again. Under a fat moon, waxing gibbous, they gave speeches about the dawning of liberty in the world. They made toasts to Washington, Jefferson, Franklin, and James K. Polk. Fiddlers fiddled. Girls and boys danced round the fires. Men shot their rifles into the night air and whooped for joy and drank still more whiskey.

Charles Stanton, camped that night with the Donners and Reeds a week's travel west of the fort, sat and watched as his fellows banged on a drum made from the skin of a dog killed the day before for the purpose. A young man camped nearby, Frank Kellogg, feasted with his companions on a pot pie made from the brisket of a large buffalo. Then they rose and unleashed a continuous volley of fire from 150 guns they had assembled around them and stood astonished as the night air crackled and popped with the sound of five or six hundred more guns answering from far up and down the valley of the North Platte.

Much farther to the west, John Frémont and the American revolutionaries who had seized Mariano Vallejo's compound at Sonoma also celebrated the Glorious Fourth. They roasted sides of beef on spits over open fires, drank fine Mexican brandy, discharged Vallejo's brass cannons, and sampled tamales provided by the villagers. In the large adobe home of Salvador Vallejo, Mariano's brother, Frémont's regular-army troops in blue uniforms along with irregulars in buckskins, an assortment of newly enlisted volunteers, and a handful of Delaware Indians who served as Frémont's scouts sang Yankee songs, drank more brandy and wine, and danced gleefully with one another deep into the night.

Mariano Vallejo spent the evening in a cell at Sutter's Fort, shivering and shaking with malaria, swatting at the mosquitoes that swarmed up from the slough behind the fort.

5

DECEPTION

On the morning of July 5, Sarah and Jay repacked their wagon, rounded up their cattle, joined the rest of the family, and moved slowly out of the cool cottonwood groves of Fort Laramie into the surrounding dry hills, bleached blond by the summer sun. They departed without most of the companions with whom they had come this far. If nothing else, Franklin Ward Graves was an independent man, and he seems to have paid little heed to the widely shared belief that for safety's sake the emigrants should always travel in large groups. When he wanted to depart, he departed. As they left the fort behind them, only two other wagons accompanied them—one belonging to the William Daniels family and the other to the John McCracken family.

To avoid crossing the river, they struggled over a steep bluff north of the fort, their wheels cutting ruts into the soft gray sandstone—ruts that would be amplified by the passage of thousands of wagon wheels in the next few years and that are still visible today near Guernsey, Wyoming. Off to the west, they could see the purple hump of Laramie Peak rising 10,240 feet on the horizon, their first good look at the Rockies. The country here was increasingly arid, the hills increasingly rocky, cut by gullies and studded with shrubby junipers and sagebrush. This was an ancient landscape, shaped only by tectonic forces and by wind and rain and thundering herds of buffalo. It was virtually untouched by the hand of man, except for the pony trails and occasional villages of Native Americans, and it had endured thus for tens of thousands of years. Sarah's generation was among the first European Americans to behold it and would also be among the last to see it in all its pristine glory, so rapid were the transformations that were coming.

Large numbers of Sioux, on their way to do battle with the Crow, continued to pass them. On the afternoon of July 6, a small band of warriors approached Mary Ann Graves, who was riding side by side with her brother Billy a bit behind the wagons and the rest of the party. Falling in beside her, the young men on their ponies studied Mary Ann carefully. Then they addressed Billy with hand signs and bits of English. They admired Mary Ann's beauty, they said, and they wanted to purchase her from her brother. Billy, whether good-humoredly or not we do not know, declined the offer. Disappointed, one of the young men laid hold of the bridle of Mary Ann's horse as if to lead her away. Billy raised his rifle and leveled it at the young man's chest. After a tense few moments, the young man released the bridle. He and his companions wheeled around on their ponies and rode off, and Billy and Mary Ann hurriedly rejoined their family.

For the next two weeks, they rolled northwest, passing and being passed by elements of what had been the Russell Party, now under the leadership of Lilburn Boggs, the fiercely anti-Mormon former governor of Missouri who had taken over leadership when Russell resigned on June 18. They slowly gained elevation as they proceeded, generally following the drainage of the North Platte but out of sight of the river most of the time. Near modern-day Casper, Wyoming, they rejoined the river, crossed to the north side, and then left it behind for good, setting a more westerly course, aiming for the drainage of the Sweetwater River, which would guide them up to the South Pass.

Among the Graves girls and the other young people moving west from Fort Laramie that July, spirits still remained mostly high. If there was little to contemplate in the increasingly desolate landscape, there was all the more time to ponder what the future held in store for them in California or Oregon. Traveling a few days ahead of Sarah, Frank Kellogg, after enjoying his brisket of buffalo on the Fourth of July and then meeting a number of eastbound travelers from California, wrote an enthusiastic letter home.

Louisa, they say chickens in California are worth 50 cents apiece and cost nothing to grow them there. So you can make your fortune there. Gesford can take them to the harbor so you will both have something to busy yourselves in this land of laziness and comfort. James, elk skins are worth $2.00 and so plentiful that a good hunter can kill 1 to 10 in a day and so come out and get rich. Emily and Polly Venable will be useful in the establishment of schools and will do well to come. Tell old Sam women are plentiful and easy to get acquainted with. Orville, they say wagons are worth 2 to 3 hundred dollars and blacksmiths are worth $5.00 a day so you and I can do well. The American girls are in great demand and the purchasers aplenty.

In short, all good folks and all good things are desirable items for trading and you had better come.

But for the older adults, particularly those with families to attend to and worry about, the visions of California and Oregon were beginning to seem like ever-receding mirages shimmering on the endless sea of sage that stretched out ahead of them. Many of them had walked the better part of a thousand miles by now. The hardships and the doubts were beginning to accumulate like the dust on the backs of their plodding oxen. The days were crackling-hot now—often over a hundred degrees in the afternoons—but because of the higher elevation the nights were frequently bitter cold. During the long hot days, they had to take care not to let their livestock drink the alkali-poisoned water that increasingly filled roadside ponds. The road itself was sandy, the kind of soft, loose footing that was hard for the oxen to pull the heavily laden wagons through. When they stopped to take their meals, yellow jackets harassed them viciously, contending with them for their food. When they walked through tall grass, ticks assaulted them. Clouds of mosquitoes plagued them in the evenings. And worse than the mosquitoes or the yellow jackets or the ticks were the swarms of buffalo gnats that descended on them whenever they stood still, swarming into their eyes, ears, noses, and mouths, biting any soft flesh they could find.*

The fraying of nerves that had begun among members of the Russell Party back in May, when they were delayed by high water on the Big Blue, slowly began to evolve into a contagion of ill will among the Boggs Party and all the emigrants up and down the trail. Families argued with other families, contending over camping spots and rights-of-way on the road. Men stood in the late-afternoon heat, covered with dust, bellowing at one another over whether to halt for the day or press on for another hour. Women, their arms and legs worn out from toting buckets of water from distant creeks and springs, their backs aching from bending over smoky campfires, eyed one another suspiciously and murmured about who was and who wasn't doing her fair share.

Among the group traveling with the Reeds and the Donners, Louis Keseberg, a tall, thirty-two-year-old German immigrant with a thin brown beard, was becoming particularly unpopular. A number of them

---

* Members of the family *Simuliidae*, these tiny, humpbacked, bloodsucking flies attack both livestock and humans. Their bites sometimes cause severe pain and swelling in sensitive humans, and badly bitten livestock sometimes die as a result of acute toxemia or anaphylactic shock. Occasionally they die of asphyxiation, literally suffocated by the masses of insects clogging their throats and nostrils.

noted that Keseberg seemed to be abusive to his pretty and petite young wife, Philippine, who had just given birth to her first son. Some said he was an outright wife beater.

By twenty-first-century standards, even the normal relations between frontier husbands and wives might well be considered abusive. Deeply sexist attitudes and assumptions governed all interactions between the two genders. The same month that Sarah approached the Rocky Mountains, her hometown newspaper, the *Illinois Gazette*, published some handy "Rules for Wives," among them these:

A good wife will always receive her husband with smiles, leaving nothing undone to render home agreeable and gratefully reciprocate kindness and attention. She will study to discover means to gratify his inclinations in regard to food and cooking; in the management of her family; in her dress, manner, and deportment. She will in everything reasonably comply with his wishes, and as far as possible, anticipate them.

These were rules that Sarah and Jay and most of their contemporaries took for granted. But there were rules for husbands, too.

A good husband will always regard his wife as his equal, treat her with kindness, respect, and attention and never address her with an air of authority as if she were, as some husbands appear to regard their wives, mere housekeepers.

Keseberg seems to have been one of those husbands who paid attention only to the first set of rules.

Increasingly, the emigrants' worries and arguments concerned grass. Grass was the gasoline of the mid-nineteenth century. It fueled the engines that propelled them forward, their oxen. As the country dried out, it was harder and harder to find sufficient grass for pasturage every night. As the grass grew sparser, it became ever more important to conserve what energy the oxen had, and the only way to do that was to lighten the loads they had to pull. All along the trail, people started to throw things overboard—things they had thought essential when they'd packed their wagons back home or in St. Joe or Independence. Among the first to go were the heaviest things—cookstoves, extra pots and pans, iron tools, and hardware. Then furniture was thrown overboard—chests of drawers, rocking chairs, bed frames, and tables. Finally, as the oxen began to heave and strain on the longest, driest hills approaching the South Pass, even smaller items had to go—extra clothes, books, linens, nonessential food items.

The sacrifices were often hard to make, not always logical, and not always voluntary. An Oregon-bound emigrant of that year, eleven-year-old Lucy Ann Henderson, watched in amused astonishment as one of the adults in her party was told he had to part with a rolling pin.

I shall never forget how that big man stood there with tears streaming down his face as he said, "Do I have to throw this away? It was my mother's. I remember she always used it to roll out her biscuits, and they were awfully good biscuits." He had to leave it, and they christened him Rolling Pin Smith, a name he carried to the day of his death.

As they approached the South Pass, the long, slow climb was tiring and the enthusiasm of the younger people finally began to wear thin. Impatient young men flailed at exhausted oxen with bullwhips, cursing them for their slowness. Adolescent girls trudging alongside the wagons slipped into sullen sulks and cast murderous looks at the parents who had gotten them into this. Toddlers, their faces flushed red with heat and exhaustion, bellowed in the backs of wagons.

A fateful question now hung in the air for many families—whether to make for California or for Oregon. Some of those who had intended to go to California had reconsidered after hearing Jim Clyman speak ill of it along the North Platte. And by now word had worked its way along the trail that the United States and Mexico were officially at war. The news injected another element of doubt and risk for many of the California-bound, but it energized others who relished nothing more than the idea of running the Mexicans out of California and taking it for themselves.

Beginning about July 11, they began to encounter an eastbound traveler traveling alone—a somewhat dubious character named Wales B. Bonney, around whom rumors of a possible murder were then swirling in Oregon, from whence he had recently and suddenly departed. Bonney was carrying an open letter from Lansford Hastings, whom he had encountered a few days earlier on the Sweetwater River. The letter was addressed to "all California emigrants now on the road." In it Hastings talked of having just explored his new cutoff. He advised all who read it to hasten on to Fort Bridger, where he would be waiting to guide them on to California via his route. Hastings, having put the letter in Bonney's hand, had retreated to Fort Bridger to await the emigrants' arrival. While he was doing so, he had ample opportunity to consult with the fort's proprietors—Jim Bridger and Louis Vasquez— about how best to advance their mutual interests.

On July 12, Charles Stanton wrote home from just east of the South Pass to report that another division had taken place in the Boggs Party, leaving only eighteen wagons California bound. That same day Wales Bonney rode into the California party's camp bearing Lansford Hastings's letter. For Stanton, who had read *The Emigrants' Guide* and written enthusiastically about it in an earlier letter home—"get it and read it. You will see some of the inducements that led me to this step"—this unexpected contact with Hastings must have seemed providential. And it must have seemed no less so for James Reed, who was increasingly worried about the lateness of the season. Hastings, the famous and well-regarded pathfinder, they now learned, was waiting for them, in person, just up the trail, to guide them the rest of the way. At the campfire that night and on subsequent nights, the discussions turned increasingly to the merits of taking Hastings up on his offer and his new route.

On July 18 the California-bound wagons traveling with the Donners and the Reeds took their noon break at the broad, bleak, windswept expanse of South Pass and then began for the first time to follow streams that flowed westward through sagebrush toward the Pacific Ocean. The next evening they camped on one of those westward-flowing streams, the Little Sandy— not much more than a wide, shallow creek running swiftly through an arid plain. A number of the Oregon-bound emigrants whom they had passed and repassed along the trail since leaving St. Joe were encamped nearby. This was the last the California-bound emigrants expected to see of them. The road forked here, the right-hand branch bearing northwest toward Fort Hall and the proven, northerly route to either California or Oregon, the left-hand branch bearing southwest toward Fort Bridger and Lansford Hastings. To this day the place is called "the Parting of the Ways." By the next morning, the California party had committed themselves to the left-hand road, and, in an act that would fuse their names with his darkly and forever in the American imagination, the men had come together to elect George Donner as their captain.

The members of the newly constituted Donner Party were a diverse lot. It was a group that was from the beginning in danger of fracturing along cultural, economic, and religious lines. There were Protestants and Catholics. There were Irish, German, English, Belgian, and Yankee individuals and families. There were men who considered themselves rich and men who owned little more than the threadbare shirts on their backs. There were well-educated men and men who likely could not even write their own names.

Most of them belonged to one or another of several large families—sixteen Donners, split between the families of George and Jacob; five Reeds, headed up by James, who was Irish born but of Polish descent; another Irish family of nine, headed up by fifty-one-year-old Patrick Breen; and six Murphys from Tennessee, a family headed up not by a man but by a widow, thirty-six-year-old Levinah Murphy. Also traveling with the Murphys were Levinah's two married daughters, Sarah Foster and Harriet Pike, along with their husbands and children.

There were smaller families, too—twenty-eight-year-old William Eddy, a cabinetmaker from Illinois, his wife, Eleanor, and their two young children; the Kesebergs with their toddler daughter and a newborn son; and another German couple, the Wolfingers.

There were also bachelors of various ages and ethnicities, among them the diminutive Charles Stanton from Chicago; John Denton, an English gunsmith from Sheffield; an elderly Belgian man named Hardcoop who traveled with the Kesebergs; Joseph Reinhardt, a German traveling with the Wolfingers; another

Irishman, a merry, brown-haired, blue-eyed friend of the Breens named Patrick Dolan; the Donners' teamsters, Samuel Shoemaker, Noah James, and a Mexican drover named Antonio whom they had hired at Fort Laramie; another German who seems to have worked for the Donners, Augustus Spitzer; three teamsters working for the Reeds, twenty-eight-year-old Milt Elliott, James Smith, and Walter Herron; and a Reed family servant, Baylis Williams, an albino who stayed in the wagon during daylight hours and did chores around camp at night. Other than the widow Levinah Murphy, the only single woman not traveling with her parents or stepparents was the Reeds' nearly deaf cook, Baylis Williams's half sister Eliza.

Despite their many differences and their frequent bickering, over the next few days the senior men of the Donner Party gradually came into agreement about one thing—the advisability of taking Hastings's new route. They were thoroughly fed up with life on the trail, and increasingly with one another. With every difficult mile they traveled, the shortest possible route to their destination seemed more attractive. There would still be an opportunity to turn north for Fort Hall when they reached Fort Bridger, but they thought less and less of that option as they moved down the road. The lyrical descriptions of California they had read in Hastings's book and elsewhere had long since begun to bewitch them, and his reassuring letter about the new route now began to seal the deal.

There was at least one dissenting opinion, but it was a woman's and therefore not likely to carry much weight. Forty-four-year-old Tamzene Donner, George's wife, was downcast and apprehensive. She could see nothing but folly in the notion of following a man they did not know over an unproven route. She could not shake her growing suspicion that Hastings might turn out to be, as she was later reported to have put it, merely another "selfish adventurer."

Sarah and her family crossed the South Pass and reached the Parting of the Ways on the Little Sandy a few days after the Donner Party. There they, too, took the left-hand road, toward Fort Bridger, parting company with the Daniels and McCracken families and continuing on alone in their three wagons. July 28 was Eleanor's fifteenth birthday, and on August 3 they arrived at Fort Bridger.

There they learned that the Donner Party had come and gone already, arriving on July 28 and leaving on July 31, hoping to catch up with Hastings, who had left days before with some sixty or more California-bound wagons. Before the Donner Party had left, they added several members to their number—another of the exceptionally tall men who characterized the 1846 emigration, a six-foot-six, thirty-year-old Kentuckian named William McCutchen, his wife, Amanda, and their infant daughter, Harriet; a sixteen-year-old French-Canadian boy named Jean Baptiste Trudeau, who claimed to be a mountain man whose father had been killed by Indians; and a young man named Luke Halloran, who was so ill with tuberculosis that his party had abandoned him but whom George and Tamzene Donner had taken into one of their own wagons simply out of compassion.

Fort Bridger as Sarah saw it was not much to behold. It was not in fact a fort at all. It was simply a small trading post run by Jim Bridger, a trapper and mountain man, and his partner, Louis Vasquez. A rough stockade of pointed poles was the only form of fortification that it boasted. When Edwin Bryant had seen it a few days before, he had found little to recommend the place. "The buildings are two or three miserable log cabins, rudely constructed and bearing but a faint resemblance to habitable houses."

Miserable it might have been, but several sinuous strands of the Blacks Fork wound their way past the fort. The streams watered lush meadows of green grass and thereby provided ample pasturage for the Graveses' livestock, which were haggard and all but broken down after the long, slow haul over the South Pass. The fort offered a last chance to rest those animals or trade them for livelier specimens, which Bridger and Vasquez had in abundance. And the establishment also offered Sarah and her family an opportunity—at exorbitant prices—to lay in a few additional provisions, new shoes or moccasins to replace their worn-out footwear, a pint or two of whiskey to replenish what had been consumed on the Fourth of July, buffalo robes with which to keep warm in the mountains ahead. The next outpost of American civilization was a very long ways down the road—a settlement called Johnson's Ranch in the western foothills of the Sierra Nevada. By the time they reached it, they would be home free, in California.

While the Graves family grazed and rested their livestock, someone—whether it was Louis Vasquez or Jim Bridger, we don't know—offered them some advice. It was the same advice that Bridger had given James Reed just four days before, advice that Reed had recorded in a letter home.

The new road, or Hastings' cut-off, leaves the Fort Hall road here, and is said to be a saving of 350 or 400 miles going to California, and a better route. . . . The rest of the Californians went the

long route, feeling afraid of Hastings' Cut-off.
Mr. Bridger informs me that the route we design
to take is a fine level road, with plenty of water and
grass. . . .

Reed seems in some ways to have been a doubtful judge of character. He had, at any rate, been mightily impressed by Bridger and Vasquez, calling them "two very excellent and accommodating gentlemen." But what Reed didn't know was that when Bridger and Vasquez had recommended the new route, they had held something essential back from him, something they didn't bother to tell Franklin Graves now either. Vasquez had in his possession letters that Edwin Bryant had left for Reed. The letters warned Reed and others following him by all means not to take the cutoff, that it was wholly impractical for wagons. But Vasquez never delivered the letters to Reed.

He had his reasons. Bridger and Vasquez had run the trading post since 1843, but by the summer of 1846 they faced an uncertain future. With most of the emigrant traffic now taking the more northerly route from the Parting of the Ways to Fort Hall, there was little reason for their establishment to continue in business—unless Lansford Hastings's cutoff, which began at their doorstep, should become the established route to

California. If that were to happen, Fort Bridger would not only survive but flourish.

Franklin Graves knew nothing of ulterior motives. He knew only what he had read and what Bridger and Vasquez had told him about the new route, and all in all it sounded promising.

While he and his family prepared for the next push forward, the Tuckers and the Ritchies and some of the other families with whom they had traveled on and off for months also arrived at the fort. All of them were planning to veer northwest for Fort Hall, bound most likely, they thought, for Oregon, and not at any rate about to hazard Hastings's unproven route. So Sarah and her siblings said their final good-byes to young George Washington Tucker, to Harriet and William Dill Ritchie, and to many of the other young people they had met back on the plains. Then Franklin Graves, Jay Fosdick, and John Snyder turned their three wagons southwest and hurried off, insofar as one is able to hurry when pulled by oxen, across still more arid, sage-covered hills in pursuit of the Donner Party and Lansford Hastings.

The Donner Party was star-crossed from the very outset. One or two days out of Fort Bridger, two thirteen-year-olds, Virginia Reed and Eddie Breen, took their saddle ponies on a cross-country gallop together. Edward's pony stepped in a prairie dog burrow and threw the boy to the ground. His left leg snapped on impact. It was a bad break, a compound fracture between the knee and the ankle. The ragged ends of the tibia protruded through bloody skin. Someone sent for help back at the fort, and many seemingly endless hours later a tattered old mountain man with a long white beard streaked with tobacco juice rode into camp astride a mule. He examined the wound.

Breaks like this were serious, life-threatening wounds in 1846. Gangrene and death were the all-too-common outcome. The mountain man unrolled a bundle of instruments and extracted a long knife and a meat saw. When Eddie saw the glint of the tools, he began to wail, and then to scream. Someone held him down.

But his parents could not bear it. They stopped the old man, paid him five dollars for his trouble, and sent him on his way, grumbling that he had not been allowed to demonstrate his skill with the knife and the meat saw. James Reed set the bone as best he could. Then the Breens laid Eddie in the back of one of their three wagons and the caravan moved on, the ends of Eddie's bones grating and grinding together as the wagon bounced over rocks and ruts.

On August 5, following the ruts left by Lansford Hastings and the emigrants he was leading, the Donner

Party passed through Echo Canyon, under spectacular red-rock ramparts that towered above them on the right-hand side of the road. On August 6 they approached the eastern end of Weber Canyon, a narrow and tortuous passage into the Wasatch Mountains. The ruts disappeared into the canyon, but before they could follow them, the Donner Party came across a note fluttering from some sagebrush. It was from Hastings, who had left it there two days before, and its contents astonished and bewildered them. It said that the road through the canyon ahead was poor, and it instructed them not to enter with their wagons but to wait there and send a messenger forward to meet him. Then he would return to show them a better route.

What Hastings had discovered within the confines of Weber Canyon was a hellish tangle of boulders, overhanging rocks, narrow ledges, and dead-end side canyons. The party he was leading had struggled for days, making an average of only a mile a day, sometimes having to all but carry their wagons over the boulders that lay in the Weber River.

In places they had resorted to using a windlass to drag wagons and teams up steep slopes. At a place called Devil's Gate, the rope hoisting one of the wagons broke near the windlass. Men rushed to support the wagon, grabbing at the spokes of the wheels and the planked sides, trying to hold it against the pull of gravity. But gravity won. The oxen bellowed and pawed frantically but futilely at the loose talus on the slope. They began to lose ground. The wagon accelerated, sliding down the slope, dragging the wide-eyed and still-bellowing oxen with it. The men had to jump free of the rig to save their lives. Then it hurtled over a precipice at the bottom of the slope, pulling the oxen over the edge two by two. Wagon and oxen cartwheeled through the air, dropped to the boulders at the bottom of the canyon, and landed in a heap of splintered wood, twisted iron, and gore.

Hastings and the remainder of the wagons made it through, eventually, but it was an experience that those who endured it would remember vividly and tell and retell when they were old.

After reading Hastings's note, James Reed, Charles Stanton, and William Pike volunteered to ride ahead and overtake Hastings. On the afternoon of August 6 or the morning of August 7, the three of them entered the narrow canyon on horseback and quickly came to see why Hastings had advised against it. Even without wagons it was a desperate scramble just to get through. It was late in the day on August 8 before they emerged tattered and torn from the mountains near the south end of the Great Salt Lake, where they finally found Lansford Hastings in the flesh.

They asked him to make good on his promise to return into the Wasatch and show them the better route
his note had spoken of. Hastings agreed to ride back
with them, but he went only as far as the top of a peak
on the western edge of the mountains. There he ges-
tured at a series of steep hills and brushy canyons that
lay between the head of Weber Canyon, where the
Donner Party was encamped, and a narrow gap just
below the summit on which he stood.

Hastings had had his fill of the Wasatch, and he still
had to guide the lead wagons across the next great obstacle presented by his cutoff, the searing salt flats of
western Utah. He also knew that the better route was
only theoretically better, and he likely did not want to
be present if and when the Donner Party found out
anything to the contrary. So he rode back down one
side of the mountain to rejoin his party while Reed
plunged down the other side into dense thickets of oak
and aspen, working his way along Indian trails, trying
to follow what he took to be the new route, using a
hatchet to mark blazes on trees as he went. Stanton and
Pike stayed at the Salt Lake for a few days to rest their
fagged-out horses.

By the time Reed struggled back into camp on
August 10, the ranks of the Donner Party had swelled
by thirteen. Sarah and Jay Fosdick, the Graves family,

and John Snyder had finally caught up with the
Donners, and with their dark destiny.

When Franklin Graves's hired man, the vibrant and
youthful John Snyder, met James F. Reed at the mouth
of Weber Canyon on August 10, he met the agent of
his doom, but he did not yet know that. That evening
Reed, despite what he had just seen in the canyons of
the Wasatch, set about convincing the rest of the party
that Lansford Hastings was right—a difficult but pass-
able route lay through the gap at the crest of the mountains. It would take considerable work, but wagons
could be gotten through by going that way. He had already named the new pass Reed's Gap, and he could
show them the way to get to it.

With this assurance from Reed, the party turned
southwest. At first the new route seemed a good one,
rising gradually through a broad canyon bordered by
more of the familiar sagebrush-covered hills through
which they had traveled for many weeks now. But
soon the canyon began to grow narrower and the climb
steeper. When they arrived at the head of the canyon,
they could, for the first time, get a good look at the
jumble of mountains that lay ahead of them. Following
Reed's blazes, they descended from the first ridge and
turned up another canyon, crossing and recrossing the

same stream, but the route soon became choked with brush.

This was not the kind of brush that one could cut through with a few lopping swings of an ax nor root out with chains dragged between teams of oxen. This was a miniature forest—ten or twelve feet tall. Along with the usual willows, cottonwoods, and aspens that grew in dense stands close to the trickles of water in the streambeds, there were large numbers of scrubby hardwood trees—mostly Gambel oak and big-tooth maple. The aridity of the summer climate, the wind that buffeted it in spring and fall, and the snowfall that weighed it down in the winter had all worked together to compress, compact, and toughen the vegetation. The oaks in particular—stout and multitrunked—proved formidable roadblocks. Teams of men with axes had to attack each one individually, and the dense, heavy wood did not yield easily to their blades.

By August 15 the wagons could go no farther. The company made an encampment. Then Franklin Graves and Jay Fosdick and Billy Graves and all the able-bodied men and boys went forward and began the herculean task of cutting a road up the last and steepest canyon toward the narrow pass above them. Progress was slow, and Reed soon grew irritated, feeling that the men were not working as hard as they ought. It took three long days of backbreaking labor before the oxen could finally drag the wagons up into the gap.

Even then their troubles were not over. The western side of the mountain was just as steep and just as densely covered in brush. They quickly found that it was nearly as hard to get heavily laden wagons down a brush-cloaked mountain as it was to get them up. It took another three days of bushwhacking and a series of harrowing descents of precipitous mountainsides before the company finally worked their way down. And then they arrived unexpectedly at one last morale-crushing obstacle—a boulder- and brush-clogged narrowing at the outlet of the final canyon leading out into the Salt Lake Valley.

Turning around was unthinkable, but they had no heart for cutting more brush and rolling more boulders out of their way. Finally they decided to assault a steep hogback ridge that lay to their southwest. One by one they yoked each wagon to long strings of oxen—as many as twenty-four at a time—and then drove them laboriously up the slope. Each time the oxen reached the top, someone had to lead them back down and do the whole thing over again. Mothers carrying infants, toddlers churning short legs, old men gasping for breath—they all struggled individually up the grass-slick hill, crawling as much as walking as the grade got steeper.

Finally, on the afternoon of August 21, Sarah, her family, and the rest of the Donner Party stood atop what is now called Donner Hill, gazing out at the Salt Lake Valley with a profound sense of relief. The landscape that lay before them looked to be lush with grass and water and certainly much flatter than the country they had just traversed. But even as they contemplated the pleasant prospect of level roads ahead, many of them also began to wrestle with a deepening sense of anxiety. Counting the time they had waited for Reed to return from his rendezvous with Hastings, it had taken them sixteen days to make just thirty-five miles through the Wasatch Mountains. They had been told it might take a week. Still ahead of them was what Hastings had told them would be a taxing but manageable forty-mile dry drive across the salt desert, followed by hundreds of miles of sage and sand hills and alkali desert before they reached the eastern face of the Sierra Nevada, still the greatest obstacle between them and California. By any means of reckoning, they were now terribly behind schedule. What exactly it would cost them was a mystery that none of them could yet divine, but it was a question that was beginning to work on them all, a worm that was burrowing its way ever deeper into their hearts and minds.

For many of them, there was now no doubt that Lansford Hastings could not be trusted. But for some of them another niggling doubt was beginning to grow. Why had James Reed—with his haughty manner, his self-assurance, his fancy family wagon, his hired hands, and his personal servants—listened to Hastings and led them into the trap from which they had just emerged? For the Graves family, for Sarah and Jay Fosdick, and for John Snyder, whose first and only knowledge of Reed was what they had witnessed in the past eleven days, the doubts were more than niggling. But they kept their thoughts about Reed to themselves, for now.

6

SALT, SAGE, AND BLOOD

After descending from the Wasatch into the Great Salt Lake Valley, Sarah and Jay and the rest of the Donner Party worked their way across the valley, trying to find dry ground on which to traverse marshes along the Jordan River. After they crossed the river, they traveled around the south end of the great lake— pale blue and frothy, its verges white with wind-whipped foam and a crust of salt.

George and Tamzene Donner's wagons lagged behind the others. They were nursing Luke Halloran, the tubercular young man whom they had taken into their wagon back at Fort Bridger. At about 4:00 P.M. on August 25, Halloran, who had come west for the sake of his health, died with his head resting in Tamzene Donner's lap. The following day the Donners caught up with the rest of the company in the Tooele Valley. There they dug a grave in the salty soil and laid Halloran's body next to that of another emigrant, John Hargrave, who had died there just two weeks before.

Death—especially before the Civil War—was an up-close and personal kind of thing in nineteenth-century America. It came visiting so frequently that no one could ignore it or hide from it. Every villager in the northern half of the country was familiar with the slow tolling of church bells, and many knew how to read in the rhythm of the bells a coded language announcing how old the victim was and of which gender. At any given time, almost any woman knew another who was dressed in formal mourning clothes, sometimes for months or years on end. Every child was familiar with the sight of somber funeral processions winding through the streets. And it was often the corpse of another child that he or she saw passing slowly by. In the first half of the century, depending partly on where they lived, and how well, between a fifth and a third of all children died before the age of ten.

Whether that of a child or that of an adult, the nineteenth-century corpse was considered a sacred object. Then, as now, it was typically treated with elaborate care and much ritual, but in the early nineteenth century it was almost always immediate family members who provided the care and performed the rituals themselves. Family members—usually female—washed and shaved the body and dressed it in a shroud. They took the body's measurements so that male family members could set about building a coffin.* When all was ready, they laid the body in the coffin and set it somewhere in the home, usually in a parlor or sitting room. When they had ice, they sometimes put it under a board beneath the body; when they didn't, they put a rag soaked in vinegar over the face, believing that this would slow decomposition. Typically, the room was decorated with black crepe, the furniture removed, the mirrors covered with white cloth. The body, once laid out, remained in the home for from one to three days, during which time friends and neighbors came to help keep watch over it, sitting by its side day and night until the time for burial arrived.

These long, communal vigils typically involved the sharing of food and drink and thus allowed the living—the family and the friends of the deceased—an opportunity to reconnect and assure themselves that their own social connections would survive the presence

---

* Many a nineteenth-century woman, knowing that death was approaching, sewed her own shroud, and many a man constructed his own coffin.

of death among them. The vigils also served a second purpose—to allow the living to make certain that the deceased was in fact deceased. Only by watching the body for several days could they reliably put to rest a fear that ran rampant in the nineteenth century—that they might bury someone alive.

When the vigil was over, the body was taken solemnly to the graveyard in a funeral procession, often with a stop along the way at a church or meetinghouse for a public funeral. Once again it was the deceased's family and friends who dealt with the needs of the body, hoisting it and carrying it, feeling the weight and heft of death on their own shoulders. And often it was they who dug the grave and then shoveled the dirt back onto the coffin, personally laying their loved one in the bosom of the earth.

Much of this intimacy with death began to diminish during the Civil War. Only commercial institutions and the government could deal with the scale of death that the war produced. Traveling embalmers followed both armies from battlefield to battlefield, making it possible in many cases to prepare the bodies of the slain for transport back home. Sometimes the dead traveled by train hundreds of miles back to the villages from which they had come; sometimes they traveled by wagon. Either way, for the first time, large numbers of bodies were washed, shaved, dressed, and laid out not by the family but by members of a newly minted profession, government-hired undertakers. The end result was that at the close of the war, by dint of sheer numbers, the corpse had begun to lose some of its sacred nature, and the handling of the dead had begun to be done mostly by strangers.

Since then the trend has accelerated, giving birth to a death industry that now hauls in as much as $24 billion a year in the United States. This industry offers the amazing gamut of products and services that anyone who has lost a loved one is aware of, among them grief counseling, refrigeration, makeup for the corpse, hairdressing, visitation rooms, recorded music, limousine service, airline shipping, and containers ranging from $39 cardboard cremation boxes to $20,000 mahogany caskets. Lately, new technologies have allowed the industry to capitalize on fresh opportunities like virtual memorial services delivered to the bereaved via streaming video, DVD remembrance montages, and a special printer ink that can be mixed with cremation ashes and then used to print out photographs of the deceased. In all of this, of course, death has become ever more abstracted, pushed ever further into the background and out of sight, something to be taken care of by those who do such things.

When Luke Halloran was buried in the salty soil of Utah, there was no close relative to wash the body and lay it out, though it is likely that the Donner women performed the task themselves. There was no bell to toll, no formal mourning clothes to be worn, no church in which to hold a service. Many of the intimate rituals of his time had to be cut short or dispensed with, but at least Luke Halloran received some sort of funeral. In the weeks and months ahead, the Donner Party would find it increasingly difficult to manage even the most rudimentary attempts at solemnizing the deaths that would occur among them.

For the next several days, the company pushed westward following the wagon ruts left by Lansford Hastings and the emigrants traveling with him, skirting low ranges of hills. Where they could find grass and sweet water, they laid in extra supplies for the hard crossing of the salt desert they knew was just ahead. Eddie Breen lay in the back of his family's wagon, the ends of his broken leg bones slowly beginning to knit.

On Monday, August 31, from a low pass in the Cedar Mountains, they saw for the first time the vast white expanse of the salt desert shimmering before them. Lansford Hastings had told them the crossing would be about forty miles, but they could see now that in fact it would be nearly twice that. To make matters worse, the last spring they had come across had been foul, the water brackish and reeking of sulfur, and they knew that out on the salt they would have to share with their cattle what water they already had stored in kegs.

They ventured out to the tabletop-flat salt desert that evening. The sun was low on the western horizon, and its light, glancing sharply off the white salt, cast improbably long shadows behind them. They kept going. There was nothing to be gained by stopping on the bone-dry sagebrush edges of the salt flats, and they wanted to travel by night as much as possible to be spared the brutal daytime heat. When the sun set at 7:03 P.M., the moon, a bit more than half full, was almost directly overhead. As the last of the sunlight faded away, the white expanses of salt beneath their feet slowly became luminous with soft moonlight. At first the landscape must have seemed almost enchanting to Sarah and her sisters. This was a pale, softly glowing world unlike anything they could have imagined. And it was strikingly silent.

They walked alongside the wagons all night. Where the salt was dry, it crunched softly under their feet; where damp with brine, it quaked softly, as if they were walking on gelatin. At 1:20 A.M. the moon slid behind the Pilot Range, still far to the west, and the pale white world quickly went black. It was hard now to keep track of their direction, to keep track of one another, and to keep track of the loose cattle that trailed behind the wagons. The heavens blossomed with stars, but the cold of the desert night deepened and tightened its grip on them. The iron-cold air was scented with brine, an odd metallic smell. They did not dare to take shelter in the wagons for fear the extra weight would exhaust their oxen, so they donned what they could in the way of cloaks or wrapped themselves in blankets and continued to walk, the miles stretching out endlessly before them in the darkness.

When the cold, pale dawn—shell pink and powder blue—finally broke at their backs a few minutes before 6:00 A.M. Tuesday, they kept on walking. From time to time, they adjusted their course to the northwest, aiming for the base of the most prominent mountain on the western horizon, 10,720-foot-tall Pilot Peak. It was their beacon. At its base, they had been told, they would find a freshwater spring. But it seemed scarcely any closer than when they'd viewed it from the Cedar Mountains the day before. They had been walking for twenty-four hours by now, and vast tracks of salt plain still lay ahead of them. Their caravan was strung out for miles.

As they walked on and the sun rose higher behind them, they narrowed their eyes to slits against the relentless glare of the light reflected off the salt. A warm wind came up and blew powdery white salt dust into their eyes. The wind itself tasted salty. Every so often they stopped and forked some grass out of the backs of their wagons and doled out water in buckets to the oxen, but before long the grass was gone and what little water was left had to be reserved for the humans.

They steered to the north of a peak rising abruptly from the floor of the salt flats. Called Floating Island, it seemed adrift on a shimmering silver sea, the mirage that lay around them now in all directions.

By midafternoon they were exhausted, weary of foot and limb, their throats parched, the mucous membranes in their nostrils desiccated, their lips cracked and tasting of salt. They were nearly blind from the glare of the sun. They at least could still take occasional sips of water. But their oxen, which had been pulling the wagons almost without rest for a day and a half now, could not, and they began to show signs of failing, still plodding forward in their yokes but pulling ever more slowly and reluctantly.

As the day wore on, both man and beast began to approach the limits of their endurance. The heat and the glare and the shimmering light started to play tricks on their eyes and on their minds. William Eddy, out front, was startled when he suddenly saw a phantom party of emigrants off in the distance, marching in lockstep with his family—men, women, horses, cattle, and dogs—all traveling in the same direction as he. When he stopped, they stopped. When he resumed, they resumed.

Elaborate mirages like this were not uncommon among those who hazarded the salt desert. Just a few weeks earlier, Edwin Bryant, standing at about this same spot in the salt desert, had seen something strikingly similar.

Diagonally in front, to the right, our course being west, there appeared the figures of a number of men and horses, some fifteen or twenty. Some of these figures were mounted and others dismounted, and appeared to be marching on foot. Their faces and the heads of the horses were turned towards us, and at first they appeared as if they were rushing down upon us. Their apparent distance, judging from the horizon, was from three to five miles. But their size was not correspondent, for they seemed nearly as large as our own bodies, and consequently were of gigantic stature. . . . I spoke to Brown, who was nearest to me, and asked him if he noticed the figures of men and horses in front? He answered that he did, and that he had observed the same appearances several times previously, but that they had disappeared, and he believed them to be optical illusions similar to the mirage.

James Reed decided to leave his family and their wagon in the care of two of his hired men—Walter Herron and his albino servant, Baylis Williams—and ride ahead alone in search of water. He instructed his teamsters to take the wagons as far as they could and when the oxen could pull no further to unyoke them and drive them forward.

Night fell again, and with it oxen began to fail all across the salt desert, sinking to their knees, bellowing, and refusing to get up. Anxiety and exhaustion evolved into fear. William and Eleanor Eddy abandoned their wagons twenty miles short of Pilot Peak and went forward with their two young children afoot, driving their cattle ahead of them in the moonlight. The Donner brothers did as Reed had done, leaving their families in their wagons, unyoking the oxen, and continuing forward unencumbered.

The Eddys made it over a low saddle between two hills and across more salt flats and finally arrived at the springs at the foot of Pilot Peak at about 10:00 A.M.

on Wednesday, September 2. Sarah, Jay, and most of the rest of the company straggled in the same morning. The Donner brothers were still far out on the desert.

James and Margret Reed and their family were in the deepest trouble, though. James reached the springs sometime on the evening of September 2, but after resting his horse and securing some water he set out back onto the salt flats about an hour later. At 11: 00 P.M. he came across his teamsters driving his oxen and cattle forward. He instructed them not to let the cattle loose once they scented water lest they stampede. Then he continued eastward. At about daylight on Thursday, September 3, he arrived back at his wagons. There he and his family sat in broiling heat under the canvas canopy of their large family wagon all day, expecting that their teamsters would arrive with water and refreshed oxen at any time. What the Reeds didn't yet know was that their teamsters had lost nearly all of their oxen during the night. By the end of the day, their drinking water was almost exhausted, so James and Margret Reed set out on foot with their children, hoping at least to reach Jacob Donner's wagons some miles ahead.

During the night a bitterly cold wind came up and the salt dust began to fly again. Finally the Reeds stopped and set their children down on a blanket and piled more blankets and shawls on top of them. When the children still complained about the cold, their parents arranged all five of the family dogs—Tyler, Barney, Trailor, Tracker, and Cash—on top of the blankets. Then the elder Reeds sat down on the windward side of the children and waited. At dawn they saw that they were not far from Jacob Donner's wagon. Elizabeth Donner took Margret Reed and her children into the wagon and tried to warm them up. James Reed pushed ahead, leaving his family behind for a second time, heading back to the springs in search of help. Later that day Jacob Donner and William Eddy returned and finally brought the remainder of the women and children still stranded on the salt flats safely in to the springs below Pilot Peak.

For nearly a week, the company stayed at the springs, bringing abandoned wagons in off the salt, resting what was left of their cattle, and desperately searching for the cattle that they had lost. Sarah and Jay and Franklin Graves had managed to save most of their livestock, so Billy Graves volunteered to help James Reed search for the eighteen oxen he was missing. The pair of them spent two long days and a night searching the salt flats in vain, much of the time chasing after rocks that the mirage effect magnified to the size and reconfigured to the shape of oxen. Finally Reed, with only one ox and one cow remaining, realized that he would have to abandon many of his possessions and ask others to carry much of the rest in their wagons. Told that she would have to leave most of her things behind, eight-year-old Patty Reed spirited away a small wooden doll, hiding it from her parents' sight lest they confiscate it. To make up a team, Reed borrowed oxen from Patrick Breen, William Pike, and Franklin Graves.

On September 10 the company awoke to the unsettling sight of snow on the higher reaches of Pilot Peak. Fourteen-year-old John Breen later remembered the reaction of some of the women in the company: "The apprehension of delay from this cause, and of scarcity, made the mothers tremble." They took an inventory of all their provisions. The results confirmed what many of them already suspected: They did not have sufficient food to last them through to California. Someone would have to go ahead to get help. As he had when someone had needed to ride ahead and overtake Lansford Hastings at the Great Salt Lake, Charles Stanton stepped forward and volunteered. So did the Kentuckian William McCutchen.

As Stanton and McCutchen set off on horseback for California, they presented a study in contrasts, Stanton at five feet five riding side by side with "Big Bill"

McCutchen at six feet six. The two men carried a letter from James Reed promising to repay Sutter for any provisions he could send back with Stanton and McCutchen.

On September 11 the rest of the company broke camp and left the Salt Lake Valley behind them. Faced with another long drive without any prospect of water, they traveled all that day and all that night. Reed's family wagon soon proved too difficult for the enfeebled oxen he had borrowed to pull, so on September 12 he stopped to again lighten the wagon by burying more of his family's remaining possessions. On September 13 the company camped near some shallow springs at a place that Reed called in his diary "Mad Woman Camp." He offered no explanation for the name except to say that "all the women were mad with anger. . . ." Whether the circle of women had hardened against the men, for getting them into this, or against one another, we do not know, but either way the bonds between and among families were rapidly fraying under the desert sun.

It's hard for us in the twenty-first century to comprehend just how squalid life on the trail in the 1840s could be. As they struggled across Nevada, Sarah and Jay lived somewhat as we now do when we go camping or hiking—except that they had no weatherproof polyester tents, no rechargeable Coleman lanterns, no flashlights, no toiletries, no propane stoves, no double-insulated iceboxes, no mosquito repellent, no subzero goose-down sleeping bags, no self-inflating air mattresses, no sunscreen, no GPS trail finders. And at four or five months minimum, it was an awfully long camping trip, at sixteen hundred miles an awfully long hike.

Maintaining any semblance of hygiene was particularly challenging. Like most people, the emigrants of the 1840s preferred to be as clean as possible. But for families like Sarah's, back home in Illinois, with no effective way to heat large quantities of water, bathing had been a seldom-indulged-in luxury. Men and women living on the frontier might take as many as one bath a week, but many took as few as a half dozen a year, and some as few as one a year. Here in the sands of eastern Nevada, bathing was a near impossibility. If it happened at all, a bath consisted of a quick, cold splash in a shallow, muddy stream, often a stream reeking of alkali or sulfur. For the most part, it just didn't happen, and so inevitably Sarah and everyone around her stank virtually all the time. They smelled not just of sweat but also of urine and excrement and menstrual blood and yeast infections and halitosis and tooth decay. Toothbrushes were a rarity, not patented in the

United States until 1857 and not mass-produced until the 1880s. As a result, even young women like Sarah often began to lose teeth in their early twenties, one reason for the stern, closed-mouth faces that look back at us from daguerreotypes taken in the 1840s. Menstrual flows were controlled ineffectively at best with rags held insecurely in place by belts around the waist.

Laundering opportunities were hard to come by as well. On the few occasions when they camped in one place for more than one night, Sarah and her sisters did as much laundry as they could. If they had time, they boiled the clothes in large kettles suspended over campfires and then spread them on rocks or hung them on the wagons to dry when the group began to move again. But as water became scarcer, the intervals between laundering opportunities grew longer. When it was necessary, the women sometimes splashed perfumes and essential oils on their bodies to mask the odors emanating from their dirty clothes.

Like their clothing, their bedding also became encrusted with dust, sweat, and the oils their bodies naturally exuded, and this tended to create fertile breeding grounds for all sorts of pests. The travelers battled body lice, head lice, bedbugs, and fleas in their wagons and tents. In the arid desert country of Utah and Nevada, their skin dried out and became scaly, their lips chapped, their eyes ached from the dust and the relentless glare of the sun. All in all, they were physically miserable much of the time.

For two weeks they traveled on, rattling through the hills of eastern Nevada, following Lansford Hastings's tracks around the south end of the Ruby Mountains on what later turned out to be an unnecessary 125-mile detour. On September 26 they reached the main fork of the Humboldt River. There they rejoined the established emigrant road, finally completing Hastings's cutoff. It had taken them sixty-eight days to reach this spot after leaving the road at the Parting of the Ways on the Little Sandy. Some of those who had stayed on the older road had made it in as little as thirty-seven days. In the end, Hastings's shortcut had added roughly a month to Sarah's journey.

They began to follow the shallow, sluggish Humboldt westward. As the days passed, they noticed increasing numbers of Shoshone Indians, short of stature, dark-skinned, and nearly naked but for breechcloths, watching them from hillsides as they passed. On September 29 two members of the Te-Moak band of Shoshones came into camp to barter and banter with the emigrants. They camped near the emigrants that night, and in the morning both they and two of the Graveses' oxen were gone. Two nights later more Shoshones spirited away one of Franklin Graves's best mares.

Since July 3, James Reed had been maintaining a daily diary that one of his fellow emigrants, Hiram Miller, had begun back in Independence in April. From the time Reed had taken over the diary, he'd mostly just noted distances the party had covered each day, road conditions, and occasional incidents of interest along the way. But on October 4 the diary concluded abruptly with one ambiguous word, "Still." It was the beginning, presumably, of an entry that Reed never finished, because of what happened the next day.

Exactly what happened in the Nevada desert on October 5, 1846, has been the subject of controversy ever since—interpreted and reinterpreted by historians and told and retold by many parties, not least of them members of the Graves and Reed families, for each of whom different truths have emerged from a welter of disputed facts. At its core, though, it was simply a nineteenth-century case of road rage.

After taking their noon break at the base of a steep, sandy incline called Pauta Pass, Franklin Graves, Jay Fosdick, and John Snyder began the arduous process of double-teaming their oxen to each of their three wagons in turn and driving them one at a time up the hill. Jay added his own team to Franklin's wagon and pulled it to the top, then returned with both teams and began to ascend with his and Sarah's wagon. John Snyder felt he could drive the third wagon up with one team and began to follow Jay up the hill. But his team of oxen became entangled with a team driven by Milt Elliott, who was struggling to make headway up the hill with Reed's family wagon. Reed, who had been out hunting with William Eddy, arrived at the scene on horseback to find Snyder quarreling with Elliott, cursing and whipping the oxen and trying to untangle the teams.

Reed dismounted. He and Snyder exchanged hot words over a wagon tongue that separated them. Snyder said Elliott had gotten in his way. Reed took Snyder to task for mistreating the oxen. Snyder raised his bullwhip and threatened to whip Reed. Reed started across the wagon tongue and drew a knife. Snyder struck him on his head with the butt of his bullwhip and then struck again. Margret Reed rushed between the two men, and one of the blows struck her on the head. Reed lunged and stabbed Snyder in the chest, puncturing his left lung. Snyder staggered a few feet before Billy Graves caught him in his arms and lowered him to the ground. And there in the Nevada desert, John Snyder, who had danced jigs on the tailgates of wagons back on the Platte and who had apparently caught the eye of Mary Ann Graves, died spewing blood into the hot sand.

Snyder had been popular, James Reed considerably less so. Reed's family and his teamsters gathered nervously around him at the bottom of the hill. Virginia Reed began dressing the wounds on her father-in-law's scalp. Sarah's family and most of the remainder of the company withdrew and encamped near the top of the hill. There they began a debate that would last through centuries about what exactly had happened. They took affidavits for a possible future trial in California, but feelings against Reed ran high, and some in the camp did not want to wait for California. Louis Keseberg propped a wagon tongue in the air and demanded that Reed be hanged from it forthwith. But some of the men gathered around Reed at the bottom of the hill— led by his teamsters—brandished rifles, and made it clear that they would fight rather than stand by and watch Reed be executed.

In the end the company decided to banish Reed with neither provisions nor weapons—the near equivalent of a death sentence in the desert, but one that would spare his wife and children the sight of their husband and father writhing at the end of a rope.

In the morning Reed offered to pry some boards from his wagon to construct a coffin for Snyder, whom he said he had always regarded as a friend. The offer was rejected, but Reed attended the funeral nonetheless. Snyder's body was lowered into the sand. Then Reed said farewell to his horrified and sobbing wife and children and set out on his prized gray mare, Glaucus, heading west alone.

That night Reed's daughter Virginia and Milt Elliott stole out of camp in the darkness, overtook Reed, and gave him his rifle, his pistols, some ammunition, and some crackers—all they could spare from the meager store of provisions they had left.

The company moved on, following the Humboldt southwest as it grew ever shallower, gradually devolving into a series of green, stagnant pools. They had expected to be in California by now, and their provisions were nearly depleted. Many families began to ration what food remained in their wagons.

Over the next few days, their fortunes continued to spiral ever more rapidly downward. Hearts that had long since begun to harden now became petrified. On October 7, Louis Keseberg put the elderly Belgian, Hardcoop, out of his wagon and told him he would have to walk, though the old man's legs had given out days before. Hardcoop quickly fell behind, but that night a party of men went back and brought him into camp. By the next morning, it had become clear that

Margret Reed's heavy family wagon, even emptied of most of its contents, was impeding the company's progress, and she was finally made to abandon it in the desert. Despite what had happened at Pauta Pass a few days before—or perhaps because of it—Franklin Graves turned one of his three wagons over to Mrs. Reed and her children.

Shortly after they got started that morning, Hardcoop hobbled up to William Eddy and said Keseberg had put him out again. Hardcoop pleaded for a ride, but everyone was walking now to spare the increasingly exhausted oxen, and Eddy told the old man he would have to walk, too. The old man limped on through the sand, but once again he quickly fell behind and was soon out of sight. That night it was bitter cold. Margret Reed, Milt Elliott, and William Eddy implored Keseberg to go back and try to find Hardcoop, but he refused. They tried Patrick Breen and Franklin Graves. Both of them had horses with which to make the attempt, but neither wanted to risk overtaxing his stock. Both refused. Some boys driving cattle into camp late that afternoon said that they had seen Hardcoop, his feet black, split, and swollen, sitting exhausted under some sagebrush some miles back. They were the last to see the old man alive.

On October 9 the group traveled all through the night. Bogged down in a long stretch of deep sand, they were unable to get free of it until 4:00 A.M. on October 10, Elizabeth Graves's forty-sixth birthday. It wasn't much of a birthday. Late in the day, Paiutes ran off all the horses Franklin Graves had tried to preserve by refusing to search for Hardcoop. The next day the Paiutes made off with eighteen head of cattle belonging to George Donner, Jacob Donner, and the German Wolfinger. Then, on the morning of October 15, while the company partook of a meager breakfast near the Humboldt Sink, the broad, marshy lake where the Humboldt River sank into the desert and disappeared, still more Paiutes crept up to unguarded cattle. This time they killed twenty-one head—all of the Eddys' team but for one ox, and all of the Wolfingers' but, again, for one. In a stroke, the Eddys and the Wolfingers were left dependent on the goodwill of the rest of the company, and goodwill was rapidly becoming a very scarce commodity.

William Eddy buried what little he had left in the way of possessions. Then he and Eleanor picked up three pounds of loaf sugar—the only food that remained to them—and set out on foot. William carried three-year-old James, and Eleanor carried their infant, Margaret.

Wolfinger, who was said to be carrying a large amount of money, wanted to bury the body of his wagon with his goods cached inside, but no one was willing to stay behind to help him except for two other emigrants of German extraction, Joseph Reinhardt and Augustus Spitzer. As the three men began to dig, Wolfinger's young wife, Doris—a tall girl who had only recently emigrated from Germany and spoke little English—went on ahead on foot with the other women. In the hills above them, the Paiutes crouched among rocks and laughed as they straggled by.

The company entered another long, dry drive—forty miles across a flat alkali desert. They traveled both day and night again, keeping only loose associations with one another now, each family looking out mostly for itself. At about 4:00 A.M. on October 16, walking under a thin crescent moon, William and Eleanor Eddy caught up with the Breen family at a group of hot springs where they had encamped briefly in the middle of the Forty Mile Desert. The boiling springs reeked of sulfur and belched plumes of steam into the night air, but the emigrants dipped water out with ladles, let it cool, and then drank it regardless of its taste.

The Breens filled their water casks and pushed on across the dark desert. The Eddys, still carrying their children, stumbled along behind them in the sagebrush.

When they paused again, Eddy asked Patrick Breen if he could have half a pint of water for his children, but Breen, with seven children of his own, refused. Desperate, Eddy seized a rifle, said he would have the water even if he had to kill to get it, and filled a bucket from Breen's cask. Breen let it go.

Sarah fared better than many of her companions for now. She and Jay still had their wagon and enough oxen to draw them, as did the rest of her family. They walked all that day across alkali flats, now and then passing columns of white steam rising from more hot springs. At sunset they kept moving, anxious to be shut of the desert as soon as possible. At about 4:00 A.M. on October 17, they encountered one last obstacle, a steep sand hill, much like the one where John Snyder had died days earlier. They double-teamed their oxen again and made the long, hard pull in the dawn light.

Later that morning they finally came to the swift, clear, sweet water of the Truckee River running down out of mountains they could not yet see—the Sierra Nevada. They knelt by its side and put their lips to the river and drank deeply and gratefully from it. William and Eleanor Eddy staggered in off the desert, and they and their children also knelt and drank, even more gratefully, from the river. William heard the joyous sound of geese cackling nearby and went off with his gun. When he returned, he had nine fat geese, the first food he and his family had had in two days, except for bits of the sugar they'd carried across the desert.

A bit later Augustus Spitzer and Joseph Reinhardt also rode into camp. But they carried sad news for young Doris Wolfinger. As they had helped her husband to bury his wagon and his other goods back at the Humboldt Sink, they said, Paiutes had attacked, killing Wolfinger and making off with all of his goods. Doris Wolfinger, they were sad to say, was a widow.

Doris Wolfinger was indeed a widow. But it would later emerge that Reinhardt and Spitzer had taken some significant liberties with the facts. It was not the Paiutes that had killed her husband. When the last of the company had moved on from the Humboldt Sink, Reinhardt had killed Wolfinger, whether in the heat of an argument or in cold blood we do not know. But with an opportunity lying before them, he and Spitzer almost certainly had begun to ransack Wolfinger's wagon, searching for the cash and valuables the Wolfingers had been said to have with them.

There were no witnesses to any of this, though, and the Paiutes made convenient scapegoats. So on October 18 the company resumed its journey, traveling up the

Truckee River now. At first they moved wearily along level but rocky benchlands on the south side of the river. Then, as the canyon narrowed, they were forced to repeatedly cross and recross the icy-cold stream, driving the wagons through the water, lurching over rocks the size of washtubs. The cottonwood trees lining the river stood like tall candle flames, brilliant with yellow leaves in the clear autumn light. Rabbit brush on the dry hillsides above the river bore trusses of equally bright yellow flowers, reminding some of the company of Scotch broom. But few of them were in the mood now to appreciate the scenery. Before them they could see dark clouds massing where they knew the mountains lay waiting for them.

As they traveled up the canyon and eyed the clouds ahead, they fretted about whether Stanton and McCutchen would ever return from Sutter's Fort with the supplies they knew they would need to make it through the mountains. And as if in answer to their prayers, the next day Stanton finally rode into camp leading a string of seven mules laden with flour, dried beef, and other provisions from Sutter's Fort. Big Bill McCutchen had fallen ill at the fort and been unable to return, but Stanton—a bachelor with no family connections in the company to draw him back—had nonetheless returned to them for a second time. Trailing behind Stanton were two young men—Miwok Indian vaqueros named Luis and Salvador, whose labor John Sutter had lent to the distressed emigrants.*

Stanton also brought surprising news. On his way back, he'd had an interesting encounter in the western foothills of the Sierra Nevada. In Bear Valley, thirty miles west of the summit, he had come across two emaciated travelers heading westbound. One of them was James Reed.

After he was banished from the Donner Party on October 6, Reed had traveled ahead quickly on horseback and overtaken the wagons of George and Jacob Donner, with whom one of his teamsters, Walter Herron, was traveling. Without mentioning John Snyder's death, nor his own near lynching, Reed told the Donners that he had been sent ahead to seek help in California. He enlisted Herron in the effort, and the two of them set off westward with only one mount— Reed's gray mare, Glaucus—and almost no provisions.

Taking turns riding on the mare, they made their way painfully across the Forty Mile Desert, up the Truckee, and into the Sierra Nevada, shooting the occasional goose, sage hen, or rabbit when they could find

---

* *Vaquero,* Spanish for "cowboy," is the origin of the English word "buckaroo."

them. As they penetrated deeper into the mountains, though, game became scarce, and they soon had nothing left to eat except for a few wild onions. Herron wanted to shoot Glaucus and eat her, but Reed would have none of that. They pressed on to Truckee Lake and over the crest of the Sierra Nevada. Herron began to grow delirious. Then Reed found a single bean lying in the dusty road. The two of them got down on hands and knees and searched for more, eventually finding a total of five beans. Herron took three of them and James Frazier Reed, until recently among the most affluent members of the Donner Party, sat down and ate the other two, his meal for the day.

The next day, October 22, they stumbled across an abandoned wagon with a tar bucket hanging from it. They scraped some rancid tallow from the bottom of it and ate that, but within minutes Reed became violently ill. Later that day, however, he recovered, and he and Herron made it down a steep descent into the upper reaches of Bear Valley, where they found a small party of emigrants talking with Charles Stanton. Reed was so gaunt and emaciated that Stanton at first did not recognize him.

Over the following days, Reed and Herron staggered on westward toward Sutter's Fort while Stanton continued eastward toward the Donner Party. Reed and

Herron came across more emigrants encamped in Bear Valley—among them the Tuckers, the Ritchies, and the Starks. These families, after parting from Sarah and her family for a final time back at Fort Bridger, had struck out for Oregon. But when they'd reached the Humboldt River, they had come across Lansford W. Hastings. Hastings, predictably, had had some advice for them.

> We met a man by the name of Hastings who advised us not to go the Oregon road, that we were nearer California than Oregon and we stood a chance of being caught in snow, . . . so we lay by a day to talk about it and think the matter over but finally concluded to take the California road.

Short of provisions and with their oxen giving out, they had barely made it across the crest of the mountains a few days before and were now resting their livestock prior to pressing on to Johnson's Ranch at the eastern edge of the Sacramento Valley. Looking back at the dark clouds gathering over the crest of the Sierra, they began to wonder whether their friends and companions from the plains would make it through. They hoped they'd had the good sense to turn around and go back down to Truckee Meadows rather than stay in the mountains. A few days later, on October 28, James Reed and Walter Herron, haggard, footsore, and almost too weak to walk, staggered up to the tall wooden gateway to Sutter's Fort.

That evening, as it began to rain hard at the fort, Reed was reunited with Edwin Bryant, who had traveled with him in the Russell Party until July 2, when Bryant had gone ahead with others on mules. Bryant filled Reed in on the state of the war against Mexico, and Reed signed a document pledging his services to Colonel John Frémont, but only after he had brought his family in from the other side of the mountains.

The next day, though he was still scarcely strong enough to walk, Reed also joined Bryant and a Reverend Dunleavy in signing another document—a petition for the rights to some land—an entire island, twenty to thirty miles long in the Sacramento River. Having arrived in California, Reed was bent on owning a piece of it.

Late on the afternoon of October 20, the lead elements of the Donner Party came around a final bend in the river and saw the expanse of Truckee Meadows—the broad valley where Reno now lies—stretched out before them, green and inviting. Beyond the meadows, they also saw, for the first time, the eastern flank of what they called the California Mountains—the Sierra Nevada—rising up above the meadows, a great granite wall, seemingly perpendicular, gray and imposing, capped with white snow, overhung by black clouds.

Over the next few days, the lead families rested at the meadows, letting their cattle graze and build strength while the families straggling behind caught up. Some of them stayed in place for three or four days. Then they began working their way up the canyon of the Truckee River into the dark mountains ahead of them.

And the mountains greeted them with a dark omen. Not at all certain that the supplies Sutter had sent with Stanton would see them through if they were delayed by weather, the men of the Donner Party decided to send a second advance party to Sutter's Fort to procure still more provisions. Two brothers-in-law, William Pike and William Foster, would ride ahead to the fort for backup provisions. As Pike and Foster prepared for the trip, though, a gun that Foster was holding discharged accidentally and the bullet entered Pike's back.

He did not die easily. His eighteen-year-old wife, Harriet, and the rest of the party gathered around him and watched in horror as he writhed on the ground, clutching at the dust and gasping for life for what seemed to those who watched to be an interminable time. Fourteen-year-old Mary Murphy later remembered, "he suffered more than tongue can tell." There was nothing anyone could do for him, though, with no doctor closer than Sutter's Fort on the far side of the mountains, and no medicine stronger than whiskey and herbal concoctions likely at hand. If he had been a horse or a dog, they might have shot him to put him out of his agony, but he had a Christian soul and could not be murdered. When he finally ceased breathing, Pike's stunned young widow was left with two daughters, an infant, and a toddler. For the third time in a month, the Donner Party paused briefly to lay a relatively young man's body in a rocky roadside grave and then move quickly on. They could not linger here with the mountains looming ahead.

As they pushed up the Truckee River Canyon, they separated into at least three groupings. The Breen family, along with the Kesebergs, the Eddys, and Patrick Dolan were out front. Sarah and Jay Fosdick, the Graveses, Margret Reed and her children, the Murphys, Charles Stanton, Luis, and Salvador formed a middle group. Margret Reed and some of her smaller children rode on the mules that Stanton had brought from Sutter's. The Donner brothers and their teamsters lagged behind the others in a third group.

The canyon grew narrow and steep-sided. They had to cross and re-cross the frigid river almost continuously now, trying to find passageways for the wagons among the boulders in the water. When they had crossed the river more than twenty times, near present-day Verdi, Nevada, they finally left the river and worked their way northwest up a dry, narrow side canyon and over a high ridge forested with large Jeffrey pines and ponderosa pines.

Survivors would later disagree about the exact dates, but on October 29 or 30 the lead group made a steep descent from the ridge into Dog Valley, a broad green vale, then climbed a second summit and descended again, this time into thick pinewoods. There, traveling through flurries of snow alternating with cold rain, they began to follow a clear mountain stream south and southwest. Late in the day on October 30, they camped in a wide grassy meadow just five or six miles short of Truckee Lake, nine or ten miles short of the mountain pass that separated them from California.

When the Breens and the other families in the lead party crawled out of their tents the next morning, a few inches of snow covered the ground. It was discouraging, but it didn't appear to be anything that would seriously impede their travel, so they moved on toward

Truckee Lake. The soft, wet snow sloshing under their feet made walking painful for many of them, so tattered and riddled with holes were their shoes, but that afternoon they emerged from the woods near the south end of the lake. There they got their first good look at what lay beyond the flat expanse of gray water.

At the far end of the lake stood a great jumble of granite cliffs, an imposing rock wall squatting squarely in their path west, rising more than eleven hundred feet above the level of the water. They had never seen anything like this in the Rockies nor in the Wasatch nor anywhere in their lives. There was a slight notch in the southern end of the wall, the pass through which they were supposed to travel, but notch or no, the thing looked utterly impassable for wagons, even under the best of circumstances. To make matters worse, every ledge and crevice and possible foothold on the face of the cliffs was already laden with deep drifts of snow.

They made their way through sparse woods along the north side of the lake and up a rough wagon road toward the granite crags ahead. But snow falls heavily in the lee of a mountain, and it had been falling here intermittently since October 7, a month earlier than usual. Over the past forty-eight hours, it had been snowing almost continuously up at the summit. Now, as they began to climb the approaches to the pass, they quickly found themselves in three or four feet of loose powder. They lost track of the wagon road but kept going anyway, trying to find their own route up the steep incline, meandering among boulders the size of houses. The snow was soon up to the oxen's chests, though, and the beasts could make no headway. The men cursed and snapped whips at the oxen, but it did no good. Finally they gave up and turned around. By evening they were back at the eastern end of the lake, where a cold rain was falling. Only a few inches of snow lay on the ground.

The Breens found a weathered shanty in which an eighteen-year-old emigrant named Moses Schallenberger had passed the winter of 1844–45 alone after having become snowbound at this same spot. All nine of them moved in for the night. The rest of the advance company crawled into the backs of their wagons and tried to sleep.

Sarah and Jay and the rest of the middle group arrived at the eastern end of the lake sometime after the Breens, most likely by the middle of the day, November 1, Mary Ann Graves's twentieth birthday. There they heard from the Breens the sickening news that they had tried but failed to make it over the pass the day before. As more families arrived, confusion and dissension gripped the company. Some felt that the rain and slush through which they'd been traveling meant that a warming trend was beginning—as it might have back home in Illinois—and that all they needed to do was bide their time until rain washed the snow from the mountains above. Others—likely including Franklin Graves, who was experienced with the ways of mountains from his boyhood in Vermont— argued that rain here simply meant more snow on the surrounding peaks and that they had not a moment to lose in getting over the crest of the mountains. They had Sutter's mules now to break a trail and Luis and Salvador to act as guides through the pass, advantages the Breens had not had the day before.

By the next morning, they had decided to try again. The Donner brothers and their retinue still had not come up to the lake, but the rest of the company set about organizing themselves for a fresh assault on the pass nonetheless. With George Donner, the captain of the party, absent, no one here had any real authority over the others. They argued about what should and should not be brought, about who could and could not ride the horses and mules, about who would lead and who would follow. Some of the men wanted to bring containers of tobacco; some of the women wanted to bring bolts of calico. Franklin Graves had to figure out what to do with the heavy hoard of silver coins he had squirreled away in his wagon. Some families opted to bring their wagons, others to leave them behind. The latter tried to pack their possessions onto the oxen, but the beasts bucked and bellowed and rubbed themselves against trees, trying to rid themselves of the unfamiliar loads. Louis Keseberg, who had injured his foot when he stepped on a sharp stick earlier in the trip, mounted a horse and tied his foot into a sling attached to the saddle. Finally, disjointed and out of sorts, they set out.

By midday they were on the approaches to the pass and the snow was up to the axles on the wagons. They sent Luis and Salvador out in front with some of the mules to break a trail, but over and over again the mules stumbled and pitched headfirst into the snow, kicking and braying loudly each time. The oxen's iron-shod feet clattered and clanged against ice and granite. The iron-rimmed wagon wheels could get little purchase on the snow-covered rocks, and the wagons began to slide backward. Jay and Sarah and the others leaned into the backs of their wagons with their shoulders, trying against all odds to push them forward. They urged their oxen on again with fresh shouts and curses and whips, but the oxen could move the wagons only a few feet with each effort. They struggled forward like this for hours, fighting their way yard by yard up the mountain. By now almost all the women were carrying children—mothers carrying their sons and daughters, sisters carrying their younger siblings.

It had not snowed for some hours, but black storm clouds had begun to pile up over the peaks immediately ahead. Still short of the summit, they came to a steep granite wall rising out of the snow. Luis and Salvador reported that they had lost track of the wagon road. Charles Stanton and one of the Miwoks went forward, skirting the cliff, to see if they could find signs of the road farther ahead. The two men made it to the pass and paused briefly to survey the flat valley and frozen lakes that lay to the west. Encouraged by the relatively easy terrain before them, they headed back downhill for the others, eager to show them the way. But by the time they returned, the company had ground to a halt.

The women who'd been carrying children in their arms through drifts up to their thighs were too exhausted to continue and had simply sat down in the snow. Some of the men had set fire to a pitch pine, and the flames had climbed up into the branches of the tree, popping and hissing. Everyone began to gather around the blazing tree for warmth. Stanton exhorted them to press on to the summit, but no one would move. Darkness was coming on quickly. People spread buffalo robes on the snow, lay down, and pulled blankets over themselves. Like everyone else, Sarah and Jay, exhausted by what they had just endured, lay down, too. They drifted toward fitful sleep in the eerie, wavering light of the burning pine snag.

A few hours later, the leading edge of a new storm slid in over the jumble of granite peaks just to the west of them. Snow began to spiral silently down out of an utterly black, featureless sky. One by one, feathery flakes landed on cold blankets and buffalo robes, on sweat-slicked hair, on shoulders turned to the sky, on soft cheeks—each flake delicate and slight, but each lending its almost imperceptible weight to the horror of what was about to happen.

PART THREE

The Meager by the Meager Were Devoured

All earth was but one thought—and that was death
Immediate and inglorious; and the pang
Of famine fed upon all entrails—men
Died, and their bones were tombless as their flesh;
The meager by the meager were devoured.

—LORD BYRON,
"DARKNESS"

COLD CALCULATIONS

On the morning of November 3, Sarah awoke to the muffled sound of someone shouting. Louis Keseberg was bellowing. He had just awakened to find himself seemingly alone on the mountain above Truckee Lake, surrounded only by soft drifts and mounds of new-fallen snow. When he had called out in alarm, the mounds had begun to move and to dissolve, gradually revealing the human beings that they concealed. Heads had popped up out of the snow all around him, like prairie dogs on a white prairie.

Sarah and Jay struggled to their feet, brushed snow from their clothing, and looked around in alarm. During the night everything had been transformed. A terrible hush had fallen over the world. Fresh snow weighed down the limbs of firs and ponderosa pines.

The granite peaks just to the west of them had become a series of sheer white walls. Everything was utterly still, except for the dizzying swirl of snowflakes still sifting down out of the slate gray sky. They looked about and saw that some of the cattle had wandered away and vanished into the surrounding whiteness.

For the first time, the anxiety that had been eating at the company for weeks gave way to something close to stark terror. They huddled around the smoking remains of the burned pine snag and tried to figure out what to do. If they tried to go forward, they would have to proceed on foot and likely flounder in the deep drifts until they died of exposure and exhaustion. If they returned to the lake, what remained of the cattle would feed them for a while, but they would face the prospect of starvation long before spring came. Neither option was good, but the latter at least offered some time to think of other alternatives. They gathered the mules and as many oxen as they could find and began to fight their way back down the mountain, wading through snow now hip-deep. All day more snow swirled down through the pine forests, covering their tracks from the day before, making it difficult to find their way. They did not arrive back at the lake until about 4:00 P.M.

At that same hour, 130 miles to the west, their old traveling companion from the plains, Edwin Bryant, who had made it through to California several weeks earlier, was in a farmhouse in the Napa Valley. He had just taken shelter from a new and particularly violent storm freshly blowing in from the Gulf of Alaska, and it made an impression on him.

> The storm soon commenced and raged and roared with a fierceness and strength rarely witnessed. The hogs and pigs came squealing about the door for admission; and the cattle and horses in the valley, terrified by the violence of elemental battle, ran backwards and forwards bellowing and snorting. In comfortable quarters, we roasted and enjoyed our bear meat and venison, and left the wind, rain, lightning, and thunder to play their pranks as best suited them, which they did all night.

At Truckee Lake that night, as the new storm that Bryant had witnessed worked its way into the Sierra Nevada, the Donner Party retreated into whatever kinds of shelter they could find. Sarah and Jay, lying in the rear of their nearly empty wagon—the wagon that had been so full of provisions back in St. Joe—huddling under buffalo robes and woolen blankets, watching the snow accumulate ever more rapidly out beyond the tailgate, must have wondered how on earth this could have come to pass, and how on earth it could possibly end well.

It snowed for eight days.

Roughly five miles to the northeast, the Donner brothers, their families, and their teamsters had also become hopelessly bogged down. Descending a steep ridge a few days before, probably on the descent into Dog Valley, one of their wagons had overturned, briefly trapping four-year-old Georgia and three-year-old Eliza in the wreckage. More seriously, the accident had broken an axle on one of the wagons. It had taken a day of work to fashion a new axle from a pine log, a day they could ill afford to lose. To make matters worse, George Donner had cut his right hand when his chisel slipped while shaping the axle. It wasn't much of a wound, but it was already starting to get infected and this made it difficult for him to use the hand.

The next day they had pushed on toward the lake in light snow. Before they'd gone far, though, they encountered messengers from the lake camp doubling back to warn them that the pass above the lake could not be crossed. The news must have stunned them. With no real alternative, they began to look for a place to make some kind of winter quarters. When they came to a wide meadow spread along a stream now called

Alder Creek, they stopped and went to work. The Donner brothers and their teamsters began to fell trees, then to buck the logs into lengths suitable for a cabin. Working with aching, freezing hands, they notched the first set of logs and used their oxen to drag them into position. They began to stack them up, building walls. By now the snow was falling hard and fast, but George Donner could work only slowly with his left hand. Jacob Donner, frail and in failing health, could do little to help. Fourteen-year-old Elitha Donner helped her father notch the logs while the other men felled more trees and hauled the logs to the site. By the time they had the first four courses of logs laid, though, the snow was falling at such a rate that it simply overwhelmed them.

Desperate now to get out of the snow, they abandoned the cabin and began to set up canvas tents in three separate camps, one for each of the brothers' families and one for the teamsters. They built brush shanties covered with pine boughs, quilts, and rubber sheets. They stacked poles against a large pine tree, tepee style, and covered them with more brush. While they worked, the younger Donner children sat on logs, bundled in blankets, the snow piling up all around them. Finally they all crawled into their miserable shelters and tried to figure out what to do next. They thought they might be able to build a real cabin once the snow stopped falling.

Over the next several days, as snow continued to drift down out of a monotonous, lead-colored sky, each of the families and individuals camped at Truckee Lake and Alder Creek sat down to make hard decisions, decisions that it now seemed clear might have life-and-death consequences. They all knew the importance of acting cooperatively; that, after all, was the essence of life in a company. But by nature most of them were independent and self-reliant. And most of them also, by now, were more or less disgusted with one another.

In the last few weeks, abrupt reversals of fortune had taken place, and the resources the company had available were now distributed in new and starkly unequal measures. At the lake, Margret Reed, probably the most affluent woman in the company back on the plains, now found herself and her children among the most impoverished. She still had her cook, Eliza, but little for the woman to cook. She had her servant, Baylis, but he was feeble and largely blind. The two of them mostly just represented more mouths for Margret Reed to feed now. The Eddys and the Murphys were similarly destitute. The Breens, on the other hand, still had most of their cattle and worldly goods.

Sarah and Jay and the rest of the Graves clan had certain advantages over some of their companions at the lake camp. They had lost every one of their horses and many of their loose cattle, but they still had most of their oxen, their household goods, a cache of silver coins, and years of experience living on the Illinois frontier under very harsh circumstances. Above all, they had one another to look out for and to provide mutual aid. Franklin and Elizabeth Graves and their children were not the kind of people to let a little snow scare them; nor would they let a little deprivation demoralize them.

At Alder Creek the Donner brothers also still had most of their possessions, including a large quantity of fine fabrics they had brought west and a considerable hoard of gold and silver coins. But unable to provide their families with any better shelter than tents and a brush shanty, their cash and their goods were of little use to them. Some of the young men who worked for them—those who had not gone ahead to the lake camp—had made a separate camp a short distance from the two Donner family camps. As single men they carried few possessions, though, and so they had virtually nothing to fall back on.

Several of the younger women—Doris Wolfinger at the Alder Creek camp as well as eighteen-year-old

Harriet Pike and twenty-three-year-old Amanda McCutchen at the lake camp—found themselves, like Margret Reed, unexpectedly without their husbands to stand by their sides. As a consequence they and their children were largely dependent on other families for food, fuel, and shelter. The unattached young men of the party similarly had nothing and no one to turn to, but at least they did not have children to worry about.

With the relentless snowfall weighing down the canvas covers on the wagons at the lake, the first priority for everyone there was to find or make some kind of more substantial, semipermanent shelter. Patrick and Margaret Breen had already taken possession of the primitive cabin that Moses Schallenberger and his companions had built in 1844, about a quarter of a mile east of the lake. Twelve by fourteen feet, with a dirt floor, it was rudely constructed of poles cut from pine saplings. A single opening served as both door and window. At one end stood a simple chimney that let escape the smoke from an open-hearth fire. Patrick Breen was not a robust man, but all he and his sons needed to do to make the cabin reasonably habitable was to stretch some canvas and hides over the roof and cover it with pine boughs. When this was done, all nine of the Breens, their friend and former neighbor Patrick Dolan, and the Mexican drover Antonio moved into the 168 square feet of living space.

Louis Keseberg, hobbled by his injured foot, could do no better than to build a simple lean-to out of poles and pine branches against the side of the Breens' cabin, and into this he, Philippine, and their two young children crawled, along with two other German members of the party, Augustus Spitzer and Charles Burger.

About 150 yards to the southwest of the Breens' cabin, near the stream that drained the lake, William Eddy and William Foster found a large boulder, nearly the size of a cabin itself. One side of the boulder was flat and nearly vertical. The men set about gathering materials to build a cabin up against the boulder. The cabin was rectangular, flat-topped, dirt-floored, about eighteen by twenty-five feet, eight or nine feet tall, built of unpeeled pine logs. The boulder at one end provided a natural hearth and chimney, as smoke could rise through a narrow gap between the cabin's roof and the face of the boulder. Into this one structure moved all six of the Murphys, all three of the Fosters, all three of the Pikes, and all four of the Eddys—sixteen people sharing 450 square feet.

Franklin Graves, characteristically, decided to build his cabin apart from the others. He selected a site nearly half a mile to the east and slightly north of the Breens' cabin, where he thought they would be more sheltered from storms but still have wood and water nearby. Working, like the others, in an almost continuous snowfall, their hands red and stiff, Franklin, Jay, Billy, Milt Elliott, Luis, and Salvador cut pine logs eight to twelve inches in diameter. Then they threw chains around the logs and used their surviving oxen to drag them to the site, notched them with axes, and began to assemble a double cabin, eight or nine feet tall. Each of the two interior chambers measured about sixteen by sixteen feet, with a chinked log wall between them. Each had its own fireplace. Like the cabins closer to the lake, this one had a flat roof of poles covered at first by canvas and then pine boughs. When William Eddy had finished working on the Murphys' cabin, he also helped Franklin complete the double cabin.

The chamber on one side was for the use of the Graves family, along with Amanda McCutchen and her infant child, Harriet. The chamber on the other side was primarily for Margret Reed, her children, Baylis and Eliza Williams, and the five family dogs. The Reeds had nowhere else to go, and Franklin Graves seems to have felt responsible for making sure that they at least had a roof over their heads.* The bachelors Charles Stanton, John Denton, and Milt Elliott would also have

---

* Not all the Reeds would later appreciate Franklin Graves's efforts, though. One surviving Reed eventually wrote tartly that Graves had built so far from the rest of the cabins "because he wished to, for he, & all of his family, had minds & wills of their own."

to squeeze in with the Reeds, as would the two Miwoks, Luis and Salvador, though all of these people at various times would also bunk in with the Graves family.

When the double cabin was finished, Sarah and her mother and her older siblings began to unpack the wagons, carrying their scant household furniture into their half of the dark, cold interior. They unpacked ceramic tableware, brass knives and forks, wooden spoons, and earthenware mugs; they set cast-iron skillets and their old Dutch oven around the fire pit; they found nooks and crannies to hold cobalt blue bottles of patent medicines, aqua-colored pickle jars, combs, mirrors, bits of jewelry, pouches of tobacco, and tin boxes containing herbal remedies. Franklin brought in his carpentry tools and drove nails into the walls to hang up their wet clothes near the fire pit. He and Jay and Billy hauled in a few old flintlock muskets and newer percussion rifles and stacked them in corners, along with an old brass pistol, boxes or bags of lead shot, black powder, flints, and percussion caps. They toted in a sack of beans that they had planned to use as seed when they reached California. Jay brought in his violin. They left their cache of silver coins outside, still hidden in the cleats in their family wagon. They cut pine boughs and arranged them on the earthen floor to serve as beds, then laid the smaller children down on them, bundled up for warmth. Then they went outside and sized up their oxen.

Hard calculations had to be made. To kill all the animals meant that they would lose the opportunity to draw the wagons over the mountains if the weather turned warm in the days ahead and melted the snow on the pass. Warming weather would also quickly spoil any meat they butchered now; there was almost no salt left with which to preserve it. On the other hand, with little more than pine branches to feed them, the already emaciated oxen would continue to dwindle in bulk if they were kept alive, their bodies offering less in the way of sustenance with every day that passed. Even as it was, their lean, stringy meat could feed the eighty-one people stranded here and at Alder Creek for only a matter of weeks.

And there was another, even tougher, piece of calculus that had to be worked through before anyone slaughtered any livestock. The Breens and the Graveses likely had a half dozen oxen each, but the Murphys had fewer, the Eddys had only one, and Margret Reed, with five children crowded into her half of the double cabin, had none. If something weren't done to equalize the situation, she and her children were going to die here, and far sooner than the rest of them.

Still, when Margret Reed approached Franklin and Elizabeth Graves and asked to buy some oxen, they must have flinched at the question. To surrender even a single animal would diminish the prospects that they, their nine children, and their son-in-law would survive what lay ahead. And they were not disposed to think well of any of the Reeds. What they viewed as John Snyder's murder was still fresh in their memories. They could not let the woman and her children starve before their eyes, though, so they sold her a pair of haggard but still-living oxen on credit, to be paid back two for one if and when they reached California. The Breens sold Margret Reed two more. For twenty-five dollars, Franklin Graves sold William Eddy one ox that had already starved to death.

Over the next several days, each family set about killing and butchering most of their remaining oxen. It was a messy business. Killing an animal as large as an ox, even an emaciated one, takes some doing. Those who had percussion rifles of a large enough caliber could try shooting the animals, aiming for the heart or the base of the brain, but it was hard to penetrate the skull or hit a vital organ. The other option was to make the killing more personal, grabbing the beasts by the horns and slitting their throats or sinking an ax head into the tops of their skulls, or simply smashing a sledgehammer into their broad foreheads.

Once the animals were dead, they commenced butchering them. Country men like Franklin Graves and Jay Fosdick knew their way around the inside of a carcass. They tied ropes around the rear legs of the oxen and hoisted them partway up into a pine tree and cut their throats to drain the blood. They slipped sharp knives under the hides to loosen sinews, then peeled the hides off the red, glistening bodies. They dragged the hides to their cabins and incorporated them into their roofing materials. Then they cut the shrunken bellies open, hacked their way into the chest cavities, cut the esophagi and diaphragms loose, and pulled out the entrails, carefully separating anything they could use from the offal they would feed to the dogs. Under the circumstances, they planned to use nearly everything themselves.

Up to their elbows in gore, they slit open stomachs for tripe. They cut out hearts, livers, kidneys, spleens, and pancreases and put them into bloody buckets. They cracked open skulls with axes and scooped out the brains. They pried open the great slobbering mouths and cut out the tongues.

They worked their way carefully and deliberately into the structure of each animal, probing with bloody, cold-numbed fingers for the openings in the joints where they could separate the parts with just a few cuts. Where they had to, they sawed through thick bones, but mostly the animals came apart easily under their expert hands. The women took the tails and got out hatchets and chopped them into short sections for making oxtail soup. Women and boys and girls stacked up the lean hindquarters and forequarters and sections of ribs and vertebrae with shreds of flesh still clinging to them, burying them in snowbanks for refrigeration.

When they were done, the snow around all three cabins was crimson. But fresh snow was still falling steadily, and it soon erased every sign of the butchery. They wiped their knives clean and put them away for now, none of them yet knowing the terrible irony that lay latent in what had just unfolded here.

As the company began to hunker down in the high Sierra, James Reed was slogging through the western foothills, trying to travel east. John Sutter had provided him with thirty horses, a mule, large amounts of flour, a hindquarter of beef, and two more Miwok vaqueros to help manage the horses. William McCutchen, who had recovered from his illness at Sutter's Fort, had joined him. Both men were determined to make it back to their families.

Two days after setting out, though, they ran into heavy rain and sleet, and by the time they got to Bear Valley, two feet of snow lay on the ground. At the head of the valley, they came across a tent in which two emigrants—Jotham Curtis and his wife—had taken refuge after crossing the mountains. Snowbound, frightened, and half starved, the Curtises were in the process of cooking the last piece of the family dog in a Dutch oven when Reed and McCutchen appeared. The two men provided Mrs. Curtis with some flour, and she set about baking bread. Then all of them sat down to a meal of bread and dog, the latter of which, after some hesitation and considerable sniffing, McCutchen tasted gingerly and finally pronounced "very good dog."

The next day, as Reed and McCutchen began to climb out of Bear Valley on horseback, the snow was thirty inches deep. As they went higher, the snow grew deeper. The Miwoks were a people of the valleys and foothills, and as afraid of Sutter as they might be, they would have none of this. That night they disappeared into the pines. In the morning Reed and McCutchen abandoned all but the nine best horses and continued. As they approached headwaters of the Yuba River, though, the horses struggled to make headway. They began to rear on their hind legs and then fall forward into drifts so deep that they were buried up to their noses.

The two men left the last of the horses mired helplessly in the snow and continued on foot. Almost immediately, though, they found themselves wallowing through snow so deep and loose that with each step they sank up to their chests. Exhausted, they finally conceded that they could go no farther. They took a last look eastward through the falling snow at the gray granite peaks that separated them from their loved ones—perhaps as close as ten or twelve miles to them at this point—and turned around.

At the lake camp, there was talk of trying again to cross the mountains. Despite all he had done in preparing a winter camp for his family, Franklin Graves had no intention of simply staying at Truckee Lake and watching them slowly starve. He, more than perhaps anyone else in the company, was determined to break out of the mountains and get help. With the Donner brothers laid up at Alder Creek, he was the oldest man in camp and, as he must have been beginning to realize, the most senior.

In many ways he was a natural to lead an escape attempt. He likely knew more about winter survival than all the other men in the camp combined. Living out of doors, or close to it, had always been his natural inclination. He was physically large and strong. He could read the weather, construct a sound shelter, and hunt with the best of them. He didn't get rattled easily. He persevered when faced with adversity. Perhaps most important, he didn't give a damn about what others thought of him.

Modern disaster psychologists have found that bold, decisive leadership greatly improves any group's ability to survive the early stages of an impending catastrophe. As floodwaters rise or a wildfire approaches, there generally is little time to waste building consensus, forging compromises, or worrying about other people's feelings. Tough decisions have to be made; bold actions have to be taken before a dangerous situation can evolve into a desperate one. From all we know about him, Franklin Graves seems to have fit the profile of just such a leader.

So when the snow finally stopped falling on November 12, Graves wasted no time in trying again to escape. The first attempt at crossing the pass had bogged down when the women carrying children could go no farther, so this time the only women who would go would be two who had no children, and whom he could trust to keep going, Sarah and Mary Ann Graves. Most of the healthy men in camp would constitute the rest of the company—Franklin himself; Jay Fosdick; William Eddy; William Foster; a few of the Donners' teamsters; two of Reed's teamsters; and Charles Stanton, Luis, and Salvador to act as guides. They improvised packs, loaded a few meager supplies on Sutter's mules, dressed in heavy layers of wool and flannel, and set off with the mules, wallowing through the snow toward the granite wall at the west end of the lake.

Even before the sun set over the pass that had been their first objective that evening, though, they staggered wearily back into camp. At the far end of the lake, they had encountered ten feet of snow, much of it fluffy powder into which they promptly sank up to their thighs. With every new step, each of them had had to pull a boot free from the snow, lift a knee up to his or her chest, swing a leg forward, shift his or her weight to the suspended leg, plunge forward a half a yard, and then repeat the whole process over and over. Even at sea level, the effort would have exhausted anyone. Here, at almost six thousand feet, it left them gasping for breath with every few steps, their hearts pounding wildly in their chests. Before they had even gotten truly under way, the snow had defeated the strongest of them.

The next day William Eddy—with the meat from his two oxen already dwindling away—borrowed William Foster's muzzle-loading rifle and resumed the long, cold hunting expeditions he had been making for some days now. There was little in the way of game, though. The local deer had all retreated to lower elevations, and the bears mostly had gone into hibernation. Thus far Eddy had been forced to settle for an owl one day, a coyote another day, and an occasional squirrel, all of which Eleanor Eddy had turned into miserable meals for their family.

On November 14, though, Eddy got lucky and came across a grizzly bear digging for roots in an exposed meadow about three miles northeast of the lake camp. He leveled his musket, took a long shot, and struck the bear. But a grizzly is a hard beast to bring down, even with high-caliber bullets fired from modern rifles, let alone a single lead ball propelled by black powder. A full-grown male can weigh as much as 600 pounds, a female as much as 350 pounds, and every ounce of them is lethal. The bear, irritated more than wounded, turned and charged. A skilled hunter, Eddy had put an extra ball in his mouth as a precaution. He quickly removed it, reloaded the barrel with powder, rammed the ball home, and stepped behind a tree. As the bear closed on him and began to round the tree, Eddy stuck the barrel of the rifle to the animal's chest and fired again. This time the bear tumbled into the snow. Eddy grabbed a stout stick, jumped on the bear, and began to beat it about the head to make sure that it was dead. That night, well after dark, Eddy and Franklin Graves dragged the carcass into camp behind a pair of the Graveses' last few living oxen.*

---

* In 1984 a team of archaeologists from the University of Nevada at Reno conducted a thorough dig at the site of the Murphy camp and found a wide variety of small artifacts, among them the bones of both oxen and a grizzly bear.

On the way back into camp with the bear, traveling through darkening pinewoods, Franklin Graves had exchanged some somber words with Eddy. He wasn't about to give up, but he now believed that he would die here in the mountains, he confided, because God would punish him for his part in exiling James Reed back in the desert and his refusal to return to search for the old man, Hardcoop, a few days later.

The Eddys shared the bear meat with the Fosters, Graveses, and Reeds, and for all of them it was a godsend. But with so many mouths and stomachs to satisfy, the bounty it provided was nearly gone within a few days.

By November 21 it had not snowed for more than a week, and bare patches of ground began to appear at the lake. It seemed likely that there would be significantly less snow at the summit than before, so Franklin Graves pushed for making another assault on the pass. This time twenty-two people would go, virtually all of the adult men and half a dozen women and older children. The greater the number who left, the fewer mouths would have to be fed at the lake and the longer their diminishing stores would last.

Again Stanton, Luis, and Salvador led the way with Sutter's seven mules to beat down a path. The long sequence of sunny days followed by cold nights had produced a cycle of thawing and freezing that had left a hard crust on the surface of the snow. This time they did not sink into the drifts so readily. They made it over the pass the first day and camped in the snow near the summit that night. The next morning, in the valley just to the west, though, they encountered deeper snow, and by midday the sun began to soften its surface. Men could still walk, more or less, on the crust, but the mules were too heavy—they broke through and sank up to their withers with each new step now. As the mules brayed and floundered in the snow, William Eddy argued for abandoning the animals and continuing, but Stanton refused. He had promised to return the mules to Sutter, and increasingly he seemed obsessed with the commitment.

Without Stanton or the Miwoks to guide them, the rest of the company knew they could not find their way down out of the mountains. Eddy offered to pay for Sutter's mules himself if need be, but still Stanton would not relent. Furious, Eddy ordered Luis and Salvador to lead the party on without Stanton and the mules, but Stanton intervened. He told the Miwoks— quite possibly correctly—that Sutter would hang them if they returned without the mules. Luis and Salvador refused to go on without the animals.

As the men argued, Sarah and Jay could look far off to the west, toward California. It was a bright, clear day. There was a cold wind at their backs. For as far as they could see under a pale blue sky, there was nothing but deep snow and the dark tops of pine trees. They turned around, facing into the bitter wind, and started hiking back over the pass and downhill into the somber shadows already falling over the lake from the surrounding peaks. So did everyone else. It was late in the afternoon before they arrived back at their cabins by the lake.

Working in a tiny notebook of his own making—just three and three-quarters by six inches—Patrick Breen had begun a diary on November 20 with the words "Came to this place on the 31st of last month . . ." On November 23, sitting in the cold gloom of his cabin, he noted the return of Sarah and Jay and the other would-be escapees, writing that "the expedition across the mountains returned after an unsuccessful attempt." And then on November 25, he looked skyward and wrote, "Cloudy looks like the eve of a snow storm."

Another low-pressure system had slid down out of the Gulf of Alaska. The next day, Thanksgiving Day, it began to snow again. This time it would go on for nearly a week, a week during which all the mules that

Stanton had refused to abandon on the summit would wander away, die, and disappear under six or seven feet of fresh snow.

Silently and implacably, serious hunger began to work its way into each of the cabins at the lake that week. Hunger is perhaps the strongest and most unyielding of human urges, according to Sharman Apt Russell, author of *Hunger: An Unnatural History.* Because it is so directly tied to our survival, it handily outcompetes most of our other emotions for our attention. It pesters us first, then nags us, and finally screams at us if we are unwilling or unable to satisfy its demands.

Deprived of food, our brains conspire with our guts to make their mutual needs our foremost and most immediate concern. Our bodies require approximately two hundred grams of glucose per day to function normally, at least half of that fueling just one particularly vital organ, the brain. When our blood-sugar levels begin to drop, our brains grow displeased—we get uncomfortable, we grow tired and irritable, we develop pounding headaches. Our stomachs also rumble in complaint. If they still find themselves empty after this complaining, they begin to produce the hunger hormone ghrelin and send it via the bloodstream to our lower brains, which promptly begin to shriek their own complaints all the more loudly.

At this point we are going to eat, or try darned hard to eat. If for some reason we can't eat, or can't eat very much, a number of other physiological and psychological processes then begin to kick in.

In a study at the University of Minnesota in 1945, a group of volunteers, all young, healthy men, underwent a yearlong experiment during which they were subjected to severe caloric restrictions. As their bodies began to react to the reduced amount of food, the young men first began to notice periods of dizziness. Then they became sensitive to cold, asking for extra blankets even on warm summer days. Their metabolisms began to slow down, their blood pressures began to drop, and their hearts began to shrink. Their lung capacity began to diminish. They generally began to lose strength and endurance. As time went on, they became possessive and defensive about whatever food they had, guarding it jealously from others. They became increasingly omnivorous, stopped disliking certain foods, and began to crave greater amounts of salt and other seasonings.

All this and more began to unfold for Sarah and Jay and their companions at the lake camp. No one was actually starving yet, for there were still small portions of lean beef to eat, meager and unpalatable as it was. The leanness of the meat, though, began to exacerbate the situation. The human body requires a certain amount of fat in order to digest and extract nutrients from meat, so even as they ate what they called their "poor beef," the Donner Party began to find that they were deriving little nutritional benefit from it.

As hunger hardened its grip on them, Billy Graves and some of his sisters hiked out onto the frozen surface of Truckee Lake, dug their way down through the snow, and sawed a foot-and-a-half square out of the hard lake ice below them. They pushed the plug of cut ice out of the way so they could lie on their bellies and peer into the dark depths. And sure enough, from time to time fat and silvery lake trout flashed by in the long column of light descending from the hole. They lowered hooks and lines and lay on the ice for hours, peering down, watching the flashes of silver, catching nothing.

Living conditions deteriorated steadily at the lake camp as the snow began to bury the cabins. The emigrants cut steps into the snow so that they could climb from their doorways up to the surface, but the interiors of their cabins grew dark and increasingly fetid. There was virtually no light except for the flickering of the fires. Smoke from the fires continuously stung the eyes of everyone who stayed inside. And even inside, even with the insulating effect of the snow piled up around the cabin, the cold gripped them without surcease. Their hands and feet ached around the clock, ached with the kind of dull, relentless pain that gets down into your bones and lives there and will not ease up.

In the Graveses' half of the double cabin, Sarah and Mary Ann and their mother crouched around their own fire, over which was suspended a Dutch oven in which they cooked small bits of the stringy beef they retrieved from the snowbanks outside. The rich but perishable organ meats and the best cuts of beef were almost certainly long gone by now, consumed within the first few days after the slaughter of the oxen. What was left was largely muscle and gristle. There was almost nothing now with which to supplement the meat, and no salt with which to season it. The more they ate of it, the more it began to taste like pasteboard to them. Sarah and her mother and Mary Ann tended, around the clock, to the younger children in the cabin, who were bored, miserably cold, and increasingly cranky. Five-year-old Franklin Jr. and the baby Elizabeth in particular wailed and whimpered.

Jay and Franklin and Billy spent much of their time out in the relentless snowfall with Stanton and Luis and Salvador, foraging for firewood. The dry pine limbs that had littered the ground when they arrived at the lake had by now disappeared under the snow, so they tried to knock dead limbs out of trees. Finally they took to felling living pines, cutting them off just above the snow line with crosscut saws, bucking them into short lengths, and splitting them with axes. But the work was exhausting, and they came back from these woodcutting expeditions cold, wet, and spent. The green wood they brought in burned poorly, filling the cabin with even more smoke.

Everyone's clothes were perpetually damp. There was no way to bathe at all now. Lice, bedbugs, and fleas continued to infest their bedding. The cabin reeked of wet wool, sweat, unwashed bodies, urine, and excrement. To relieve themselves, day or night, Sarah and her family had either to use chamber pots or to emerge from the cabin into the bitter cold outside, climb up steep steps cut into the snow, walk a reasonable distance from the cabin, and squat in the cold, stinging snow.

On the other side of the log partition that separated the two halves of the cabin, just eight or nine inches away, Margret Reed, her albino servant Baylis Williams, and his sister Eliza labored under similarly miserable conditions to provide for the Reed children. Margret Reed was carefully rationing her very limited supply of meat, and Baylis, who was likely getting the least share, was beginning to grow noticeably weak. To pass the time, Patty Reed played with the small wooden doll she had hidden away back on the western edge of the salt desert.

At the Murphy cabin, where there were seventeen mouths to feed, only scraps of beef were left from the two oxen they had started with at the beginning of the month. William Eddy went out to hunt every day but seldom returned now with anything more than an occasional squirrel. In the Breen cabin, Patrick was suffering bouts of agonizing pains from kidney stones. On November 29 he and his sons managed to kill the last of their oxen, but his wife, Peggy, had to do most of the butchering the next day. On December 1, as the snowstorm continued unabated, Patrick Breen wrote in his diary, "Difficult to get wood no going from the house completely housed up. . . . The horses & Stanton's mules gone & cattle suppose lost in the snow no hopes of finding them alive."

Things were even worse for those at the Donner brothers' camps on Alder Creek. Their tents and brush shanties did almost nothing to keep out the cold and snow. Their clothes were wet day and night. Most of their cattle had been lost in the first storm and buried under the snow. By now George Donner's cut hand had become badly infected, throbbing with pain and swelling up to twice its normal size. The infection was beginning to creep up his arm. When he went foraging for firewood, he had to carry it back bit by bit cradled in the crook of his right arm. Jacob Donner, frail to begin with, had also begun to weaken noticeably and now spent most of the time prostrate in his tent. Tamzene and Elizabeth Donner rationed out meager bits of beef to their children.

By far the worst off, though, were the single men living at Alder Creek—particularly James Smith, Samuel Shoemaker, and Joseph Reinhardt. Living under a brush cover in the snow with no oxen, no resources of their own, and asked to do much of the heavy work for the ailing Donner brothers, they had begun to sink from the outset. By early December they'd been reduced to catching mice with their hands, roasting the tiny bodies over a fire, and consuming them whole. When that fare proved inadequate, they had begun to cut strips from their buffalo robes and eat them.

At the lake camp, the failed attempts at crossing the mountains began to exact a noticeable physical toll on those who had made the effort. The laws of supply and demand were starting to catch up with them, particularly the strongest and most energetic among them.

It is possible, under certain circumstances, to live for a very long time with very little food, or even with no food. Christians, Muslims, Buddhists, Jews, Hindus, and American Indians—just to name the most obvious examples—have all at various times in history embraced extended fasting as a means of attaining heightened levels of spirituality. Saints of various sorts have fasted for months. In the nineteenth century, so-called hunger artists often fasted for twenty, thirty, or in one case forty-four days without apparent permanent harm to themselves. In 1981 the IRA's twenty-seven-year-old Bobby Sands survived 66 days on a hunger strike before he died. Another twenty-seven-year-old man in Scotland survived an astonishing 382 days on nothing more than water, potassium, and sodium sup-plements. His trick: He started his fast weighing 456 pounds. Paradoxically, though, you can starve to death in as little as two to three weeks. It all depends on the math.

How many calories a particular individual needs to consume depends on numerous variables, the most im-portant being his or her age, height, weight, and degree of activity or inactivity. Nutritionists typically use a formula known as the Harris-Benedict equation to fig-ure out how many calories a subject needs to consume simply to maintain his or her current weight. The equa-tion produces a number called the basal metabolic rate, or BMR. When converted from metric to English units of measure, it looks like this for women:

$$BMR = 655 + (4.35 \times \text{weight in pounds}) + (4.7 \times \text{height in inches}) - (4.7 \times \text{age in years})$$

The equation is interesting because by making a couple of educated guesses about Sarah's weight and height, we can use it to figure out roughly what her basic caloric needs were when she became snowbound in November 1846. Both Sarah's mother and father were notably tall and thin. If we assume conservatively that she was something like five feet eight inches tall and weighed perhaps 125 pounds, the formula tells us that Sarah would have required about 1,612 calories per day. But that assumes she was lying in bed, night and day, expending no more energy than required to eat, breathe, think, maintain a core body temperature, and carry on the other business of keeping her body functioning.

To figure out anyone's true caloric requirements, the Harris-Benedict equation requires one further step—assessing that person's level of activity and then multiplying the BMR by a factor corresponding to that level. If we peg Sarah as a "very active" woman, as she certainly was during those frantic weeks in November and December of 1846, we must multiply her BMR by 1.9. This yields an estimate that Sarah required about 3,063 calories per day—roughly five and a half Big Macs—simply to maintain her weight. By way of comparison, in 2007 the average American woman consumed an average of 2,679 calories per day.*

---

* Up sharply from 2,158 calories in 1970.

If we run the similar but slightly different Harris-Benedict equation for males and make similar estimates about Franklin Graves's height, weight, and age, we discover that Sarah's father needed to be taking in something like 3,646 calories a day, considerably above what the average American male now eats. These figures, though, probably understate by a good deal the number of calories that Sarah and her father were burning, for two reasons: First, they assume a person of average fitness and with an average amount of muscle mass. By the time Sarah and her father had walked much of the way across the continent, they likely were far more fit than is now average and had acquired very high percentages of muscle mass. The greater the amount of muscle, the higher the caloric demands of the body. Second, the calculations don't take into account the often bitterly cold environment in which Sarah, her father, and everyone in the Donner Party were operating. In those kinds of conditions, even lying in bed, the body requires far more calories than normal.

The net result was that for some time now they had been burning calories a good deal more rapidly than they were taking them on board with their daily rations of lean beef. All their bodies could do in response was to quietly and efficiently begin to cannibalize themselves in order to provide energy to the brain and other vital organs. Sarah and Jay and their companions discovered that they were beginning to feel weak. They began to grow gaunt. Their eyes began to sink deeper into their faces. Their fingers grew bony. Ribs and other bones began to protrude in ways they had not previously. And as all these transformations took place, they began to peer into one another's increasingly angular faces with a growing sense of alarm and incredulity.

8

DESPERATION

In the first two weeks of December, Franklin Graves was still determined to make a break for it, despite the string of earlier failures. He knew that the most recent report that anyone in California had heard regarding their situation was whatever James Reed and Walter Herron, who had left the company far back in the Nevada desert in early October, might have told them. That meant that for all John Sutter or anyone else in California knew, the company might at this moment be wintering in reasonable comfort in Truckee Meadows, with access to plenty of game and plenty of grass to keep the oxen alive. There was, therefore, no particular reason to believe that anyone would come looking for them before spring. If even a few of them could get through, somebody might be able to keep what seemed about to happen here from actually happening.

So within the close confines of their cabin, the Graves family set up a manufactory for snowshoes, something that Franklin had used as a boy in Vermont and that he hoped would be the means of their salvation. With help from Charles Stanton and others, they dug through the snow searching for abandoned oxbows—the U-shaped pieces of bentwood that fit under the necks of the oxen and connected them to their yokes. Franklin carefully split these lengthwise along the grain of the wood to produce from each one a matched pair of thinner but still-substantial bows. Sarah and her siblings cut long, narrow strips of rawhide from the skins of the slaughtered oxen and wove them together tightly in a crisscross pattern over the frames provided by the bows. When they added wider rawhide straps to hold their feet in place, they had durable, if heavy and cumbersome, snowshoes, each about two feet long and a foot wide. By early December they had fifteen pairs of them stacked in the cabin.

On December 9 one of the Donners' teamsters, Augustus Spitzer, left the Keseberg shanty, where he had been staying. Like many of the single men, he had few resources to fall back on, and it is unlikely that the Kesebergs had much that they were willing to share. Apparently nearing starvation, Spitzer staggered around the corner of the Breens' shanty to their entrance, descended the snow steps, and collapsed full length through the doorway. Patrick and Margaret Breen dragged him the rest of the way into the dark cabin, another mouth to feed.

On that same day, Charles Stanton sent a note to the Donners at Alder Creek.

9th Dec 1846 [Mrs. Donner,] You will please send me 1# your best tobacco. The storm prevented us from getting over the mountains we are now getting snow shoes ready to go on foot I should like to get your pocket compass as the snow is so very deep & in the event of a storm it would be invaluable Milt & Mr. Graves are coming right back and either can bring it back to you The mules are all strayed off—If any should come round your camp— let some of our Company know it the first opportunity Yours Very Respectfully C.T. Stanton

Stanton remained an essential element of any plan for escape. He was the only one in the company who had crossed the summit and returned, and everyone counted on him and the Miwoks to find the way. But if he was going up that mountain yet again, he aimed to make sure that he had not only a compass with which to navigate, with or without the Miwoks, but also sufficient tobacco for his pipe. If he made it through to California for a second time, he intended to stay put this time. But Franklin Graves, he knew, would be coming back for the rest of his family as quickly as possible.

Over the next few days, Franklin Graves made his way laboriously from shanty to shanty, once again recruiting the youngest and strongest of the company for the attempt. At the overcrowded cabin built against a boulder, he asked Levinah Murphy, "Are there any in your cabin, Mrs. Murphy, that want to go? It is our only choice." It was an agonizing decision for the thirty-six-year-old widow and her family, but they finally settled on Levinah's married daughters, Sarah Foster and the newly widowed Harriet Pike. The two young mothers—both of whose breast milk had by now dried up—would leave their babies behind in Levinah's care. Sarah's husband, William Foster, would go, too. So would thirteen-year-old Lemuel Murphy and ten-year-old William Murphy. William was of such a slight build that they thought perhaps he could walk in the footsteps of the others, without snowshoes. William Eddy would go as well, but Eleanor would stay behind to care for their two small children.

At the relatively well-stocked Breen cabin, only the lighthearted bachelor Patrick Dolan elected to go. He was the one single man who owned enough beef to almost certainly survive the winter on his own, but he insisted that Margret Reed and her children should have some of it, the rest to go to the Breens. In the Kesebergs' lean-to, Charles Burger, the Donners' teamster, thought he would go, too, though he would attempt to do it without snowshoes.

At the Graves-Reed double cabin, the decisions were simpler. Of Margret Reed's four children, only Virginia was old enough to even be considered, but she had fallen ill in recent days, and it had become clear that she could not go. Salvador, Luis, and Charles Stanton had to go—they were the guides. Antonio, the Mexican drover, would go.

On the other side of the log partition, the Graves family made its own decisions. Elizabeth would stay behind to care for her children, with help from twelve-year-old Lovina and fourteen-year-old Eleanor. Billy would also stay behind, to chop wood, tote water, shovel snow, and take care of the other heavy chores for his mother and younger siblings. Sarah and Jay and Mary Ann were young, strong, and vigorous. All would go. Mary Ann convinced Amanda McCutchen that she should go, even though it meant leaving her baby, Harriet, behind for Elizabeth Graves to care for. And, of course, the man to whom all eyes had now turned, Franklin Graves, would also go.

If anyone started out with a particular disadvantage, though, it was Franklin, for all his practical knowledge. At fifty-seven, he was more than twenty years older than the next-oldest men in the party, Patrick Dolan and Charles Stanton. In the long run, when stamina became the difference between life and death, the age gap might well prove telling. But Franklin Graves's children were on the verge of starving, and he did not intend to let them down.

They all knew that this would be the final attempt, that everyone's lives now hung in the balance, and that the odds were heavily weighted against them. No one would be turning around this time. There was no reason to come back without provisions, and plenty of reasons not to. Returning empty-handed would only mean starving and watching one's family starve. They also knew that anyone who could not keep up would have to be left behind to die a cold and lonely death.

They bided their time, watching the weather, waiting for the right moment to make their break for the pass. By December 12 it had been snowing again for four days straight. The next day it continued, more heavily than before. Patrick Breen watched the snow mount around his shanty and observed in his diary that it "snows faster than any previous day. . . . Stanton &

Graves with Several others making preparations to cross the Mountains On Snow shoes, snow 8 feet deep on the level. . . ."

On December 14 the day finally broke clear and fine and rather warm, but the snowshoe party stayed at the lake. Hunger was their constant companion now, gnawing at them from the time they awoke every day until they fell asleep at night, and it urged them to do something, anything, as soon as possible. But with so much fresh powder sitting loose on the surface, they feared that they would quickly get bogged down, even on snowshoes. So they waited. On the fifteenth, conditions were the same: clear and dry and relatively mild. The powder was still fluffy. That night, though, there was a change. The air was iron cold, and that was what they had been waiting for.*

On the morning of December 16, the day once again dawned clear, but now, finally, there was a firm crust on the surface of the snow. In the Graves cabin, Sarah and Mary Ann donned heavy flannel pantaloons, garments that they had likely contrived themselves by altering their heavier dresses. They put on linen shirts, woolen coats, and cloaks. They pulled on woolen socks

---

* They may also have been waiting for Milt Elliott to return from Alder Creek with the compass Stanton had requested from the Donners in his letter of December 9.

and battered boots. Franklin and Jay dressed in woolen trousers, woolen shirts, and woolen hats. Jay wrapped a black scarf around his neck. Those without scarves wrapped rags around their necks, anything to keep the cold at bay. Then they put makeshift packs on their backs. The packs contained blankets, a little coffee and sugar, some tobacco, and about eight pounds of dried beef for each of them.* This, they thought, if they rationed it carefully, would see them through for the six days that they calculated it would take them to make it through to Johnson's Ranch in the western foothills. They believed that Johnson's was thirty or forty miles to their southwest. In fact it was sixty-six as the crow flies, at least seventy-five by the route they would attempt.

Finally they clambered out of their cabin and stood outside in the snow, bending over in the bright sunlight, strapping their new snowshoes to their boots. Now all that remained was to say farewells that they well knew might be their last. Amanda McCutchen had to give her baby a final kiss. Sarah and Mary Ann had to embrace their mother and their younger siblings one more time. Franklin Graves had to look Elizabeth

---

* A pound of beef jerky—essentially the type of beef they carried— yields an average of about 1,208 calories, according to modern packaging labels.

in the eyes and tell her he would return. And then the moment inevitably came when those who were going had to turn their backs on those who were staying and begin to make their way off through the snowy woods.

It could not have been easy, but they were hastened on their journey by a sobering fact lying in the snow nearby. Earlier that morning, or the night before, Billy Graves had washed and shaved the cold, stiff body of Baylis Williams. Then he and John Denton had dragged Baylis out of Margret Reed's half of the cabin, cut through the hard crust of snow, and buried him six feet deep in the softer snow underneath. The dying that they had been anticipating and dreading for weeks now had begun.

They made their way to the Breens' shanty where they met up with the other snowshoers. More tearful partings took place. Harriet Pike had been as reluctant as Amanda McCutchen to leave her infant daughters behind, but unable to nurse her one-year-old, Catherine, she believed that the best thing she could do was to go for help. Yet she was in anguish now that the moment had come. So were Sarah and William Foster, leaving behind two-year-old Jeremiah George. Perhaps no parting was more difficult, though, than that between Eleanor and William Eddy. The two had been struggling together simply to keep their children alive, almost continuously since Paiutes had killed all their oxen in the desert two months before. Now, as William finally turned his back on Eleanor and his children and began to walk away, he was racked by silent, tearless sobbing.

The snowshoe party—what the historian Charles McGlashan would later call "the Forlorn Hope"—struck off through pinewoods toward the eastern end of Truckee Lake, wallowing in the snow as much as walking on it. Most of them had never used snowshoes, and they had a hard time getting the hang of it. Even with the frozen crust, their feet sank a foot or more into the snow with each step. With each subsequent step, they had to pull the bulky snowshoes free from the holes into which they had sunk on the previous step. They grunted and gasped at the labor of it. They fell forward and backward and sideways into the powder, trying to move forward with some degree of control and efficiency. Gradually they learned to manage the cumbersome snowshoes more effectively, but the effort required to use them was rapidly exhausting them even as they set out, and burning what would turn out to be precious calories at a furious rate.

Two of their number—Charles Burger and ten-year-old William Murphy—had no snowshoes, and they were having an even harder time of it. Young Murphy quickly found himself up to his thighs. Burger, known by all as "Dutch Charley," was short and stocky. His stout legs punctured the surface of the snow like pile drivers, and he had to try to bull his way forward by brute force, up nearly to his hips in snow.

They moved out onto Truckee Lake. The sky overhead was bright blue, and there was a light, cool breeze at their backs. By midmorning the sun warmed them, but its light reflected harshly off the white expanse of snow and ice covering the lake. As the sun moved farther to the west, out over the jumble of granite peaks at the far end of the lake, the glare began to strike them full in the face, dazzling their eyes and threatening to blind them, though they did not yet know that.

If you ski or snowboard or climb mountains, you likely have at least a passing awareness of snow blindness, though these days any reasonably good pair of sunglasses will protect you from its effects. Called radiation keratitis by medical professionals, snow blindness is caused by exposure of the eye to ultraviolet B rays (UVB). Unlike ultraviolet C, the most dangerous form of ultraviolet radiation, UVB rays are able to penetrate the ozone layer in the earth's atmosphere, though their strength is much diminished by its filtering effects. Under normal circumstances, at sea level, the eye can absorb UVB rays without damage. But with every thousand feet in elevation gain, the strength of ultraviolet rays increases by 5 percent. So at the elevation of Donner Lake, for instance, UVB rays are approximately 30 percent stronger than at sea level. The exposure of the eyes to UVB rays is greatly increased by the reflectivity of snow, and a snowy environment at high altitude is where one is most likely to suffer from snow blindness.

Under these circumstances, particularly after prolonged exposure, UVB rays irritate the superficial epithelium of the cornea. This produces an inflammatory response that results in symptoms ranging from mild irritation to acute pain, nausea, headache, temporary loss of vision, and even permanent blindness if the exposure continues long enough.

Snow blindness can be particularly insidious for those who are unaware of its dangers. Symptoms often lag the actual exposure to UVB by as much as six to twelve hours, so one may go blithely about his or her business unaware of the damage that is being done until it is too late.

There is one tender mercy that snow blindness offers to most of its victims. The cornea is remarkably good at healing itself. Given a chance, it will repair the damage spontaneously within about forty-eight hours.

That, however, benefits only those who are able to get in out of the snow and sunlight or take other precautions. Those who continue to expose themselves day after day to the same levels of UVB cannot heal. Their eyes inevitably suffer something like what a sunbather's skin would suffer if he or she persisted in lying out in the sun every day with red, blistered skin upturned to the sun's unrelenting rays.

As the day wore on and the snow covering Truckee Lake grew softer, Charley Burger and William Murphy gave up and turned around partway across the lake. Without snowshoes they simply could not keep up with the others, and with the cliffs ahead growing closer, they must have known they would never make it over the pass. That left fifteen in the snowshoe party—nine men, five women, and a boy. Beyond the lake there was another mile or so of gradually rising, forested, and boulder-strewn terrain before they would reach the abrupt granite cliffs that they would ultimately have to climb to reach the pass. The days were short now in mid-December, and they needed time to make camp before the sun set. So when they finally trudged off the ice at the western end of the lake, they climbed a small hill and removed their snowshoes and packs. They hunted for firewood, gathered pine boughs for makeshift beds, struck a fire using sparks from the flintlock rifle that Eddy was carrying, built a fire on a platform of logs, spread out their blankets, and sat down to ration out their first portions of dried beef.

As Sarah sat with Jay on their pile of pine branches, chewing the cardboardlike beef, she could look down the length of the lake at the semicircle of snowy trees fringing the far end, painted peach and rose by the light of the sunset streaming over the pass above them. In among the trees, blue curlicues of smoke rose from the cabins. The easternmost of those, she knew, was from the fire her mother was tending. At this very moment, her brothers and sisters were gathered around that fire eating their own dwindling share of the beef, wondering about her as much as she was wondering about them. It was a hard, cold thing to be so near and yet so irretrievably far.

She and Jay slept that night, or tried to sleep, beneath thin blankets under clear, crystalline skies. The frigid black vault of the heavens above them was moonless but ablaze with shimmering stars. The southeast wind that had blown all day continued through the night, making the long, low, mournful sound that only wind in pines can make.

The next day, December 17, began with what they believed would be the toughest challenge they would face—climbing the east face of the pass. The weather was bright, clear, and cold again, all they could have hoped for in that regard. They made a little coffee, chewed a bit of dried beef, strapped on the snowshoes, and began climbing through thickly forested country toward the granite cliffs that lay beyond.

At first the going was relatively easy—over the first mile they gained only about 250 feet in elevation. But that modest rise brought them out of the forest and into a very different kind of landscape, one in which there were few trees but large expanses of granite rising abruptly out of the snow. Like much of the crest of the Sierra Nevada, this landscape had been scoured by a series of glaciers over the past million years. Moving ponderously downslope, sculpting out the depression in which Donner Lake now lies, and then retreating again, these glaciers had left behind a jumble of highly polished flat surfaces, house-size boulders, deep crevices, and abrupt cliffs. After the last of the glaciers finally melted away, another ten thousand years or so of additional exposure to brutal weather had shattered much of the granite into loose talus—piles of broken rock that had accumulated in deep drifts on some of the gentler slopes. To all this was now added perhaps ten or twelve feet of snow and liberal applications of ice wherever water had run over the granite and then refrozen.

They were only three-quarters of a mile from the narrow notch in the summit that constituted the pass itself, but they still had nearly a thousand feet to climb before they reached it. They worked their way up among the rocks, following Salvador and Luis. With the snowshoes strapped to their feet, they could not walk easily on rock and ice, so they tried to stay on open expanses of snow, but at intervals they were forced to cross stretches of slippery rock. As the grade steepened and the air grew thinner, they had to stop more and more frequently, gasping for breath before pushing on.

A month earlier all of them had likely been in far better than average physical condition. Most of them were young. Their systems were by now acclimated to the 5,936-foot elevation of the lake camp. They had for the most part walked all the way from the Missouri River, climbed over the Wasatch, cut brush, toted water, and chopped firewood for weeks on end. If they had been well nourished, their aerobic capacity should have been nearly optimal. But they were far from optimally fit. Their meager diets had by now begun to erode both their muscle mass and the capacity of their lungs.

The powdery snow and steepness of the climb made every step harder than they might have expected when they set out. The snow deepened toward the summit.

Even with their snowshoes, the men in particular found themselves sinking deeply with each step. Sarah and the four other women—of whom twenty-three-year-old Amanda McCutchen was the oldest—moved out in front. Because they were lighter, the women were better able to tackle the powder and beat down a path. The men began to follow in their footsteps.

By early afternoon they were high among the cliffs, scrabbling for footholds and handholds now as they worked their way up the ever-steeper route. The bright sun bore down on them through a relatively thin atmosphere at nearly seven thousand feet, and the reflection of the sunlight off of snow and rock and ice began to take an additional toll on their bodies. Charles Stanton, in particular, began to feel the effects of snow blindness.

All of them were also likely suffering in another way from the effects of the sun. Thus far they had managed to stay dry enough and warm enough to avoid what inevitably would be the greatest danger they would face in their quest to escape the mountains—hypothermia. But as they struggled up the mountain wearing layers of heavy, dark wool, the glacier-polished granite, ice, and snow began to reflect the sun's heat from every surface, and their bodies increasingly had to labor to ward off hypothermia's opposite—hyperthermia.

. . .

Ordinarily the human body is quite adept at maintaining a steady core temperature very near 98 degrees Fahrenheit. When our core temperatures begin to vary even by a degree or two from this fixed point, our bodies take measures to return themselves to a state of thermal homeostasis, a normal temperature. When we get too cold, for instance, we shiver as the body vibrates small muscles around vital organs in order to burn calories and generate warmth. Goose bumps rise on the skin in an effort (mostly vain in our case) to fluff up our primordial fur and provide an insulating layer of air above our skin. When we get too warm, on the other hand, we sweat so that evaporation can cool the surface of the skin, carrying heat away from the core of the body. But the margins for error are narrow. Hyperthermia begins to set in when core body temperatures rise above 101 degrees Fahrenheit. Brain death begins at 106 degrees. Conversely, hypothermia begins when core body temperatures sink below 95 degrees. Death occurs at 86 degrees. So whether the outside temperature is 110 degrees as we trek across the salt flats of Utah or 10 degrees as we sleep in the snow on Donner Pass, our bodies must maintain their inner workings within about a 6-degree range if we are to remain reasonably functional, and within a 20-degree range if we are to remain alive.*

In a 1982 experiment conducted on climbers ascending Alaska's Mount Denali and documented in the PBS television show *Deadly Ascent,* Dr. Peter Hackett discovered that even well-conditioned alpine climbers sometimes experience dangerous fluctuations of internal body temperatures. Using data transmitted by NASA-designed radio-thermometer capsules that each climber swallowed, he found that the exertion of climbing on sunny slopes while dressed in clothing designed to retain body heat could cause core temperatures to soar rapidly into the hyperthermic range. Even more troubling, he found that those same climbers' core temperatures could plunge just as rapidly into the hypothermic range when they stopped climbing and sat down on snow or cold rock. Aside from the direct dangers posed by hyperthermia and hypothermia, Hackett's study suggests that the temperature fluctuations that occur within the bodies of climbers place enormous physiological stresses on the body as it struggles mightily to return itself to the state of thermal homeostasis on which it depends for survival. This additional stress, added to all the other stresses of climbing—the thin

---

* Interestingly, whole-body hyperthermia is now sometimes deliberately induced in clinical settings to weaken and damage cancer cells.

air, the extreme exertion, the unrelenting need for concentration, the glare of the sun, the threat of frostbite, and so on—is sometimes what pushes climbers' bodies over the edge, into a kind of death spiral.

As Sarah and Jay struggled up the face of what would eventually be called Donner Pass on December 17, 1846, they knew nothing of hypothermia or hyperthermia. But under all the layers of sweat-soaked wool in which they were swaddled, their bodies were already fighting a silent, internal war between death by fire and death by ice, swinging back and forth between thermal extremes in a way that threatened to destabilize their regulatory systems and their bodies' precious reserves.

By late afternoon they had scaled the pass and stood near the eastern end of the long valley in which Sutter's mules had become bogged down back on November 22. There—despite their exhaustion and the desperate situation in which they found themselves—a few of them paused for a few moments to take in the grandeur of the scenery. A cluster of snowcapped peaks lay just to the southwest. To the north stood the massive basalt buttresses of 9,104-foot-tall Castle Peak. The sky overhead was a pale, translucent blue. The darker blue of the lake glinted far below them like a polished oval of lapis lazuli. Someone commented that they were about as near to heaven as they could get. Mary Ann Graves stopped to watch her companions move out in front of her and took note of her surroundings: "The scenery was too grand for me to pass without notice. . . . Being a little in the rear of the party, I had a chance to observe the company ahead, trudging along with packs on their backs. It reminded me of some Norwegian fur company among the icebergs." Not everyone was taking in the sights, though. Most of the others simply pushed doggedly on ahead, heads down, taking one heavy, awkward step at a time.

None of them moved far beyond the pass that day, though. The hours were too short, the snow too heavy, their bodies too exhausted. Just west of the summit, they once again built a fire on a platform of logs, chewed their meager rations of beef, and fell asleep on the snow. Despite their exertions they had traveled only about three miles from the previous night's camp.

The third day, December 18, once again dawned clear and cold. They slogged the length of the valley, through deep powder. When they left the valley behind, they moved southwest, skirting a high granite peak and then descending very slowly toward a cluster of frozen lakes. The sun remained bright, the snow a brilliant white, and the temperatures moderate through the morning, but in the afternoon the sky clouded up and snow flurries started to blow into their faces out of the northwest. By late afternoon Charles Stanton began to lag far behind the others.

Stanton had several strikes against him from the beginning of the escape attempt. His compact, five-foot-five-inch frame made it exceptionally difficult for him to manage the heavy snowshoes as well as his longer-legged companions did. He, more than any of the others, had been battling snow blindness since the first day out, and he was by now likely feeling its full effects—nausea, headaches, and, of course, the loss of much of his vision. Like most of the other single men, he had been living on exceptionally short rations for weeks before they'd left the lake, surviving essentially on any extras that the families with oxen had been able to share with him. And perhaps most important, largely because of his magnanimity, he had by now already hiked across the crest of the Sierra Nevada three times, not to mention the treks across Wasatch and the salt flats and the Forty Mile Desert. Now, finally, he was beginning to reach the limits of his endurance.

Sarah and her companions reached a cluster of conifers in late afternoon and again prepared a fire, but it was another hour before Stanton finally staggered into camp. His declining state was a cause of much anxiety among his companions, not only because he was almost universally liked and admired but also because they all knew he was their most reliable guide. Although Luis and Salvador had also traveled over the pass, eastbound, there had been no snow on the ground then, and the landscape was now much altered. They had never gone in this direction, as Stanton had, and they spoke little English at any rate. If the party lost Stanton, they knew they themselves might quickly be lost.

Beginning at about 11:00 P.M., it began to snow again. Sarah and Jay shivered and quaked through a more uncomfortable night than any they had thus far experienced.

On the morning of December 19, there were still intermittent snow squalls, but as the day wore on, the skies began to clear and the party once again found themselves moving through a bright white, blinding landscape. To their south, steep and heavily glaciated granite peaks rose abruptly out of the surrounding forest. A little to their north lay the Yuba River, tunneling deep under the snow in some places, breaking free and tumbling brilliantly in sunlight among snow-frosted boulders in other places. Sarah and her companions trudged on almost due west all day, following the sun. A number of them, among them Mary Ann Graves, were now experiencing varying degrees of snow blindness. Once again, though, it was Charles

Stanton on whom the sunlight was taking the greatest toll, and it was he who began to fall the farthest behind.

They were now into the fourth day of what they had thought might be a six-day journey, but they were only about fourteen miles west of the lake camps. They had just two more days' rations of beef, and Johnson's Ranch was still more than fifty miles to their west. Late in the afternoon, they stumbled down into deep blue shadows lying along the frigid bottomlands beside the Yuba River and made camp. Once again it was an hour or more before a dazed and exhausted Charles Stanton finally trudged into camp.

Sunday, December 20, the fifth day out, began ominously. Though the skies overhead were again clear, dark clouds were gathering on the western horizon, far out over the Sacramento Valley toward which they all yearned. As Jay and the rest of the party began to move down the Yuba, Sarah and Mary Ann hung back, trying to fix a problem with Mary Ann's snowshoes. When they had made the repair, Sarah started on ahead to catch up with Jay. As Mary Ann began to leave, though, she noticed that Charles Stanton had not departed with the others. He was in fact sitting quietly nearby, resting his head against a snowbank, puffing on his pipe, making no effort to get going. Mary Ann asked if he was coming along, and he said yes, he would join them shortly. She hesitated. Stanton gazed in her direction, but his snow blindness had severely damaged his vision, and he likely could not see her. He did not get up. Finally Mary Ann turned and hurried away to catch up with the others. Stanton continued to sit there smoking. Five months later a party traveling eastward found his bones in a hollow stump near the same spot.

If a modern coroner had conducted an autopsy on Charles Stanton's body, she likely would have found that—weakened by long-term hunger and malnutrition—he'd died of hypothermia as a result of sitting still in the bitter cold. Chances are his core body temperature simply dropped gradually below the 90-degree-Fahrenheit threshold at which shivering stops and the final symptoms of hypothermia begin to kick in—amnesia, clumsiness, difficulty speaking, mental confusion. At about 86 degrees, his skin would have begun to turn blue, his respiration and pulse would have slowed, major organs would have begun to fail, and finally his brain would have died.

But Charles Stanton probably died psychologically before he died physiologically. As John Leach points out in *Survival Psychology*, science has long recognized that under some circumstances people are able "to die gently, and often suddenly, through no organic cause." In other words, we are able, sometimes, to will ourselves to death, or at least to cease willing ourselves to live.

In 1972, in a situation much like that which Stanton, Sarah, and the other members of the snowshoe party faced, Nando Parrado had a kind of epiphany, as he relates in *Miracle in the Andes*. It came to him in his moment of maximum despair. He and his fellow rugby players, trying to hike out of the location where their airplane had crashed high in the Andes, scaled a steep ridge that they thought separated them from safety, only to find that snowy peaks stretched away from them in all directions. Aghast at their predicament, Parrado fell to his knees in the snow and took in a staggering realization. Death was the rule, life the exception. Life was at best a transitory dream, set in a universe that was entirely indifferent to his fate. Whether to cling to that fragile dream, Parrado realized then and there, was up to him as it is up to all of us, moment by moment. Whether to embrace what we are all thrust into, squealing with astonishment and rage, or to fall back into the comfortable, dark, quiet realm of the insentient. Nando Parrado decided to fight for the dream. Charles Stanton, it appears, after all his heroic efforts to aid his fellow travelers, had chosen to slip back into the darkness.

. . .

The snowshoe party traveled down the Yuba toward an abrupt granite knob now known as Cisco Butte. There they left the river, turning south to climb and cross the eastern flank of another peak, then turning west again. From time to time, they paused at high spots and scanned the snowy landscape to their rear, looking for Stanton. They were increasingly apprehensive about losing him, but they knew they could not afford to stop so long as the weather held out. Each time, failing to see him, they pressed on into the still and silent whiteness. They crossed bridges of snow twelve or fifteen feet thick arching gracefully over streams. They passed places where springs lay twenty or twenty-five feet deep at the bottoms of wells that the water had melted in the snow. In the afternoon they finally descended into a flat piece of terrain now called Sixmile Valley. It had been a hard day, nearly eight miles of tough snowshoeing, and it had exacted a heavy price from them.

They had by now spent four nights lying out in the open in the high Sierra, where nighttime temperatures in December run in the low twenties or high teens. They had not yet even begun to descend below the snow line, but their supply of dried beef was already nearly exhausted. Hunger cramped their stomachs and clouded their thinking. Their boots—soaked and tattered before they had even started—had now begun to fall to pieces, and as a result their feet ached from continual exposure to snow and ice. The men in general seemed to be faring worse than the women. Sarah's father in particular was growing weaker, and even Jay was having a hard time keeping up with Sarah. She began to hang back with him, traveling a bit in the rear of the rest of the party.

When they made camp that night, setting fire to a dry tree and gathering around it as had become their habit, they wondered again, more urgently now, how they would find their way forward from here without Stanton and whether "here" was in fact where they were supposed to be. But they had bigger problems than they yet knew. As they sat around the burning tree that night, changes were under way high in the upper atmosphere, changes that had been months in the making.

In certain years, years when La Niña conditions prevail in the southern Pacific, a meteorological phenomenon known as the Madden-Julian oscillation is sometimes born in the Indian Ocean. The MJO, as meteorologists call it for short, carries vast amounts of relatively warm and very wet air from the Indian Ocean into the central

Pacific. When cold low-pressure systems move south out of the Arctic, as they regularly do, they siphon this wet air northward from the central Pacific, drawing it toward the Pacific Northwest. If the cold air from the Arctic collides with the wet air from the central Pacific, enormous amounts of precipitation fall along the West Coast of the United States. Those of us who live in the Northwest usually get the brunt of this phenomenon, which we not so lovingly refer to as "the Pineapple Express," but sometimes the storm track sags to the south, carrying the moisture-rich air into California. The result of such scenarios is typically widespread flooding in the lowlands and, sometimes, epic blizzards in the high Sierra.

One such record-breaking storm hit this section of the central Sierra Nevada in January of 1952, dumping more than twelve feet of snow on Donner Summit within a few days. While it was just one in a series of storms that produced a total of sixty-five feet of snowfall that year, this particular storm generated news around the country. On January 13, enormous wind-blown drifts trapped a westbound passenger train, the City of San Francisco, at Yuba Gap. At first the 226 passengers and crew made light of the situation. Many of them were servicemen on their way to San Francisco to be shipped off to the Korean War, and they had no objection at all to sitting in one place for a while, eating the railroad's food and drinking its liquor. But after thirty-six hours the diesel fuel ran out and temperatures in the passenger cars began to plummet. Food began to run short, and tempers began to flare.

Hundreds of volunteers worked with shovels to reach the train, but it took them seventy-two long hours to get there. As the passengers finally hiked out along the path that had been cleared through the drifts for them, many of the children wore on their heads pillowcases with eyeholes cut in them to protect them from the frigid winds still slicing through the mountains. No one knows if any of the passengers thought about the Donner Party as they sat on the train for those three days, but they might well have—the spot where the City of San Francisco was stranded at Yuba Gap was not more than a half a mile from where Sarah and the snowshoe party likely camped on December 20, 1846.

At the lake camp, on opposite sides of the log partition that separated the Graves side of the cabin from the Reed side, Margret Reed and to a lesser extent Elizabeth Graves were engaged in preserving the lives of their children. To stretch the little beef they had left, to make it last as long as possible, the two mothers had begun to cut strips from the hides of the oxen that had been slaughtered. They held the strips over open flames, singeing off the hair, then boiled them until the collagen separated from the hides and formed a thick, unpalatable, but reasonably nutritious glue. This glue they rationed out to their children, who gagged it down unhappily.

Things were even harder for the souls huddled miserably in wet tents and under brush shelters at the Donners' Alder Creek camp. The robust young men who had wrangled cattle, cut brush, driven oxen, and rolled boulders out of the paths of wagons for the Donners and Reeds for months had begun to die, unable to subsist on a diet of roasted mice and strips of toasted buffalo robe. James Smith, Sam Shoemaker, and Joseph Reinhardt were all dead. As he lay dying, Reinhardt had confessed to Doris Wolfinger that he had murdered her husband back at the Humboldt Sink.

Jacob Donner was also dead. He had been the first of them to die. Never very robust, he had descended into a state of nearly complete inaction almost as soon as they became entrapped. For weeks he had dwelled in despair, doing little to help himself or his family, until finally one day he sat down at a table in the tent, bowed his head upon his hands, and sat motionless until he died.

9

CHRISTMAS FEASTS

B y the morning of the snowshoe party's sixth day out, December 21, it had been snowing hard through much of the night. As daylight arrived, snow continued to fall, shrinking their world down to a circle twenty or thirty yards in diameter. Beyond that distance everything faded into a blur of white snow, gray rock, and dark green conifers. Without Stanton to guide them, they turned to Luis and Salvador, but the young Miwok men were no more able than the rest of them to see beyond the white curtain enveloping them all. Finally, with no other real choice, they strapped their snowshoes back on and set out again.

They managed to hold a generally westward course through the morning. There was a strong, relatively warm wind out of the southwest. From time to time, the snowfall lightened a bit. During one of these intervals, Mary Ann Graves, looking down into a deep gorge to the north, believed she saw smoke hanging in the air. She began to holler at the top of her voice, but there was no answering cry. She prevailed upon the men to fire the flintlock rifle, but it drew no response. She implored the men to turn to the north and descend into the gorge to investigate the smoke, but Luis, who spoke a bit of English, said it was not the right direction, so they trudged on toward what they believed was the west.

Before they had gone much farther, though, they again ground to a halt. They stood in the falling snow talking, arguing. Their stock of dried beef was nearly gone; they did not know where they were, nor even with any certainty which direction they were going. The parents among them desperately wanted to hold their children again. Some of them argued for turning around, but it had taken them six days to get this far, with provisions. With no food to sustain them, attempting to return seemed suicidal, though going forward seemed no less so. Mary Ann Graves said she would rather die than return and watch her brothers and sisters starve at the lake. Luis and Salvador outright refused to go back. The two Miwoks turned and resumed walking. Mary Ann followed them. Then

Sarah Foster fell in behind. Then everyone else did as well.

Sometime that afternoon they made a catastrophic mistake. As they left the western end of Sixmile Valley, they approached a low ridge to their northwest. If they had climbed it, they would have found themselves precisely where they needed to be, on the established emigrant road at Emigrant Gap at the point where it dropped some seven hundred feet into Bear Valley. From there they would have had a relatively easy, gradual descent to Johnson's Ranch. But the ridge screened their view of Bear Valley, and instead of ascending it they turned left, to the south, skirting the ridge and beginning to follow terrain that led inexorably and invitingly downhill.

Immediately ahead of them now was the canyon of the North Fork of the American River, other than the Yosemite Valley perhaps the most dramatic feature of the Sierra Nevada's western flank. A steep-sided fissure carved out eons ago by glaciers and by the river that tumbles among granite boulders at its bottom, the canyon is, for much of its length, more than three thousand feet deep, in places four thousand feet. To this day, stretches of it are inaccessible except by river rafts and helicopters. The deep side canyons that run into it are similarly steep-sided and impressive. It is a place of breathtaking beauty both in summer and in winter, but for anyone on foot, particularly in winter, it can be a world of pain and desperation at best, a death trap at worst.

But they were traveling blind; they had no idea what lay ahead. They forged on for only another mile or two, following the path of least resistance, down along one of several ridges paralleling one of the American River's tributaries, the North Fork of the North Fork. They didn't travel much farther that day, though. It was the shortest day of the year. Before the sun set at 4:39 P.M., they set another dead pine afire and made another miserable camp in the snow.*

Sarah and Jay wrapped themselves in blankets and stared into the flames. They chewed on the last few shreds of dried beef from their packs. Occasionally, as the fire climbed higher into the tree, flaming limbs broke off and plunged to the snow, landing among them, sputtering and hissing. They were so exhausted and dispirited that they made no effort to move out of range of the falling firebrands. William Eddy dug

---

* They didn't know it, but the river in the canyon to their south ran, many miles downstream, directly past Sutter's Fort. Had they had a modern inflatable raft and the skill to navigate Class-5 rapids, they could have been at the fort within a day or two.

into his pack to find something and came across a small parcel wrapped in paper. Written on the parcel was a simple message: "Your own dear Eleanor." Inside was about a half pound of bear meat. As Eddy ate, the rest of them began to ponder what kinds of choices they were about to face.

It snowed all night again, and on the morning of December 22 a few intermittent flurries were still falling. The snowshoers again shoved their bruised and aching feet into cold, wet, and increasingly tattered boots, strapped on their snowshoes, and set forth. But a warm southwest wind had come up, and it made for softer, wetter snow, which clung to the snowshoes in heavy clumps that rendered it nearly impossible to make headway. Within a short time of setting out, they struggled miserably back to their campsite and resolved to spend the day there. They gathered firewood and tried to build a new fire, but the snow was honeycombed now with rivulets of water. Every fire they managed to start simply sank into the mush and was promptly extinguished.

None of them, except for Eddy, had had anything to eat that morning. As the day wore on, their blood-sugar levels began to drop, making them anxious and irritable. Hunger pains gnawed at their guts even more ferociously than ever. Their heads pounded. Their bodies were beginning to burn protein instead of glycogen for fuel, accelerating the process of wasting that had been slowly resculpting their bodies for weeks. Their heart rates increased, their blood pressures fell. They were increasingly clumsy, inclined to fall down and disinclined to exert themselves. Their cognitive abilities were also beginning to decline—their alertness, concentration, and ability to focus on a task were all failing them.

So they sat motionless in the snow, their bodies slowly losing heat. The afternoon rapidly dissolved into another cold, wet night. In the dark they shivered and shook under their blankets, each of their bodies starting to fight a renewed and more desperate battle against hypothermia.

Just how insidious hypothermia is, and how long the odds against Sarah and the snowshoe party were now growing, is underscored by a tragedy that unfolded almost exactly 160 years later. On November 25, 2006, thirty-five-year-old James Kim and his wife, Kati, and their two daughters found themselves snowbound in their Saab station wagon after making a wrong turn onto a logging road in Oregon's Coast Range.

For more than a week, the Kims remained in the cramped, cold confines of their car. They ate berries and rationed a small supply of baby food and crackers. When the food was exhausted, Kati Kim breast-fed both her infant, Sabine, and her four-year-old, Penelope. James Kim ran the car's engine at intervals in order to provide heat, until the gas ran out. Then he removed the car's tires and burned them one by one. They huddled around them for warmth in the stench of burning rubber and waited for someone to find them. And in fact rescuers were beginning to close in on them, using signals from the Kims' cell phone to get a fix on their general location. But the Kims did not know that.

After more than a week in the car, James Kim, like Franklin Graves long before him, decided he had to get help for his family. He studied an Oregon state road map and concluded, incorrectly, that the town of Galice was just five miles away. Early on the morning of December 2, he built a final fire for his family and then set off on foot with his map in hand, telling Kati that he would be back by 1:00 P.M. He was dressed in extra layers of clothes, but he wore only tennis shoes on his feet.

Two days later a helicopter search team spotted Kati waving an umbrella just as she was herself setting off from the car with her children in search of help. James had not returned. Kati and the kids were promptly rescued, and after a night in the hospital they were fine, other than for some minor frostbite on Kati's toes. James Kim, meanwhile, had been waging a desperate battle against the elements.

He had headed south and west at first, traveling about three miles before he entered the drainage of Big Windy Creek. He then apparently decided that the creek would lead him to Galice, or at least to some form of civilization. He followed the creek back eastward, in the general direction of the car, through rugged, steep terrain broken by narrow ravines and abrupt cliffs. He tore off pieces of the road map and dropped them along the way, presumably to mark his trail for searchers, or for himself if he decided to reverse his course.

As Sarah and her companions had been 160 years earlier in the Sierra Nevada, Kim was weak from days of near starvation. And, as in the Sierra, nighttime temperatures in the Coast Range were well below freezing, and not much above it during the days. He had no shelter. His feet were wet from snowmelt and from wading back and forth across the creek. Then James Kim began to remove his clothes.

Forensic pathologists call it "paradoxical undressing." In addition to the disorientation, mental confusion, and cognitive challenges that come along with the final stages of hypothermia, many victims experience, toward the end, a sudden, overwhelming sensation of warmth. In earlier stages of hypothermia, the blood vessels of the extremities and the skin constrict in order to shunt blood and warmth to the core of the body. In the final stages, though, the process is often reversed, as vessels in the extremities—deprived of vital glucose and energy—give up the ghost and relax. Blood begins to flow rapidly away from the body's core, back out to the face and extremities, and the victim suddenly, and paradoxically, feels flushed and warm even as he or she freezes to death.

Toward the end James Kim started to shed layers of clothes, dropping them along the way. Then he lay or fell down on his back in the icy waters of Big Windy Creek and died. He never knew that he had circled back close to the car, never knew that Kati, Sabine, and Penelope would survive. But he had done what countless mothers and fathers have done through time—reached deep inside himself, marshaled all his energy, exercised his wits, and finally hazarded everything for the sakes of those whom he loved.

All fourteen of the surviving snowshoers made it through the night of December 22. On the morning of December 23, their eighth day out, they set off again, moving in single file downhill. It was a bit warmer now, and snow flurries began to give way to cold showers of rain. The terrain grew steeper, falling away off to their south as they traveled on. This country was nothing like what Stanton had told them to expect. They had been looking for a sharp drop-off to the northwest, with a flat-floored, oval valley at the base. Salvador and Luis were clearly as bewildered as the rest of them were.

At some point that afternoon, they stopped to rest. They sat in the snow, leaning against trees, studying one another's faces for signs of hope but finding only despair. They began to ask one another, "What will we do? What *can* we do?" The questions, and the answer that many of them had likely already begun to contemplate, hung darkly in the air for a long while. Finally Patrick Dolan, the merry Irishman who had sacrificed all his beef for the women and children back at the lake camp, gave terrible weight and form to it. The men must cast lots, he said, to see which among them should die to provide flesh for the others.

It was an appalling solution, and in some ways a surprisingly premature one. Though they had been on scant rations for a week, they had been entirely out of food for only hours. Most, if not all, of them were to a greater or lesser extent Christians. Dolan himself, an Irishman, was likely Catholic. And even without religious considerations, the moral imperatives against taking another life, let alone the almost universal taboo against consuming the victim's flesh, were powerful inducements to continue suffering rather than to take such a step. And yet they sat in the snow and discussed it, and the proposition began to make headway.

William Foster would have none of it, but Dolan and the other men persisted and finally carried the day. They tore up strips of paper, and the men somberly took turns drawing them. Sarah and Mary Ann had to sit and watch with dread as Franklin and Jay took their turns. Sarah Foster did the same as her husband took his chance. But it was Patrick Dolan himself who drew the fatal strip.

Dolan didn't have long to react. The men gathered around him and studied him, then looked one another in the eyes and realized that none of them was willing to put the flintlock rifle to the young man's head and pull the trigger. Dire though their straits were, murder was still murder, and, for now at least, a gnawing stomach could not supersede their moral codes, nor simple human compassion. William Eddy pointed out that one or another of them was bound to die pretty soon anyway and that they could then decide whether or not to consume the body.

They pushed on and made eight miles that day, camping somewhere on the northern flanks of the canyon of the American River's North Fork. They were utterly exhausted. They had not eaten in forty-eight hours now, but they were burning calories as if they were competing in an Olympic biathlon competition. And all the while they were sliding deeper into a topographical funnel.

From the time they had departed St. Joe until they arrived at Truckee Lake, Sarah and her companions had had one principal aid to navigation—the tracks and ruts left by hundreds of wagons that had preceded them west that summer. But once the snows of late October covered the tracks leading over the Sierra Nevada, they had in a sense been blinded. In an age when maps of the West consisted of little more than pencil sketches drawn from the recollections of trappers and explorers, and the navigational advice dispensed in guides like Lansford Hastings's was often equally sketchy, they'd had nothing more than Stanton's, Luis's, and Salvador's memories to guide them forward. Before they left the lake camp, Stanton had requested the loan of a compass from the Donners, but Milt Elliott had apparently not returned from Alder Creek in time to hand it to Stanton before they set off on their snowshoes.

It is difficult for us, with our twenty-first-century view of the earth—replete with satellite imagery, the Internet, twenty-four-hour news broadcasts, GPS systems, and high-resolution topographical maps—to comprehend just how potentially terrifying it was for the snowshoe party to come to any fork in what they imagined to be their route to salvation. They knew all too well that a wrong turn, any wrong turn, could mean the difference between living and dying. And the landscape was beginning to tell them that they had in fact already made just such a wrong turn, but they didn't know where or when. All they knew was that they were now profoundly lost, and it was beginning to eat at their minds.

By the morning of Christmas Eve, a hard, steady, cold rain had set in over the western flanks of the Sierra Nevada. During the night it had soaked Sarah and her companions, chilling them, bringing on spasms of shivers, and setting their bones to aching. They managed to build a smoky fire, but they had unknowingly camped above a snow-covered stream. The fire melted through the snow and suddenly dropped into the stream. They crawled to the hole it left behind and peered down into the dark void. They could hear the icy water running below.

As the day wore on, it grew colder. Relatively warm, wet, subtropical air that had been flowing into the Sierra from the southwest had begun to collide with an Arctic low-pressure system and the colder air that it had brought down from the Gulf of Alaska. The rain turned to sleet. Everyone in the company was ashen-faced now, but the faces of some of the men especially began to take on an almost blue pallor. Franklin Graves shook violently and incessantly in his wet clothes. Young Lemuel Murphy began to rant and rave incoherently. So did Patrick Dolan.

Both Dolan and Lemuel Murphy were likely suffering from a toxic combination of woods shock, hypothermia, and hunger-induced psychological stress. But another factor might also have been at work within Dolan's increasingly stressed psyche. Just twenty-four hours before, he had drawn the lot that told him he was to be murdered, a piece of information that could only have induced enormous psychological trauma. In a similar incident in 1765, sailors aboard a storm-damaged ship called the *Peggy* in the Atlantic, facing starvation after eighteen days without food, killed a black slave and ate him. Several days later they were ready to kill again, but no more slaves were available. So they drew lots. A foremast man named David Flatt drew the fatal lot. Flatt asked that the execution be postponed until the next day, but during the night he grew first deaf and then delirious. The next morning the crew was rescued, but by then it was too late for David Flatt. He had become permanently deranged.

The snowshoers made no effort to leave the camp. It was clear by now that at least some of them would never see California. The question was whether any of them would see it, whether someone would die quickly enough to save the rest.

It continued to grow colder. Lying on the snow under wet blankets, shuddering convulsively with the cold, men and women alike began to cry out in anguish, to no one in particular or to God, begging for deliverance, for food, for warmth. Harriet Pike found a small patch of her cloak that was still dry inside, between the shoulders, and pulled out bits of raw cotton batting. With shaking hands, using sparks from Eddy's flintlock rifle and the cotton as tinder, the men finally managed to get a fire going. But when they went to chop more wood for the fire, the head of the hatchet flew off the handle and was immediately lost deep in the snow.

Antonio, the young Mexican drover, crawled over to the fire and lay down. After a while his hand fell into the fire pit, and he did not remove it. Someone pulled it away, but the second time it fell into the fire pit, nobody bothered to move it. He was dead.

The storm intensified and the temperature dropped sharply as night came on. The wind picked up, and the tops of the pine trees around the party began to tilt over to the northeast. It started to snow hard, extinguishing the fire. By 10:00 P.M. the wind was howling through the trees, blowing the snow horizontally, plastering it against tree trunks and boulders. Franklin Graves stopped shivering. His face was blue, his pupils dilated, his limbs rigid, his breathing shallow. Eddy crawled over to him, looked him in the face, and told him he was dying. Graves said that he did not care. But he called out for his daughters.

Distraught, Sarah and Mary Ann sat by their father weeping. They hugged him and pulled him close and chafed his limbs and tried to warm him. He spoke to them slowly and weakly, his speech slurred by hypothermia. He said that their mother's life and the lives of their brothers and sisters depended on their making it through the mountains to get help. He pleaded with them to do whatever it took to survive. He told them his body must be used for food, and that they, too, must eat human flesh. Then he turned to the eighteen-year-old widow, Harriet Pike, who also sat by his side, and reminded her of her babies, Naomi and Catherine, back at the lake.

At about 11:00 P.M., Franklin Ward Graves died in the driving snow, with his daughters at his side.

For Sarah, as for Mary Ann, the devastation must have been nearly complete. Her father had been among the most hale and hearty of men and, adult though she now was, her stalwart protector since childhood. With him now dead, lying stiffly out in the open, snow already beginning to cover his blue-white face, she and her sister faced two cruel possibilities—dying miserably, as their father had, or following his dying wishes. It was a hideous choice. And even if they followed his wishes, they knew it might stave off death for only a short time. Sarah had one consolation that Mary Ann did not, though. When she retreated, sobbing, from her father's body, she had the arms of a loving husband in which to shelter, at least for as long as he lived.

Without more substantial shelter, though, more of them were likely to die before the night was out. The snow was falling even more heavily now, slanting through the air on the bitter-cold wind. It frosted the men's beards and clung to the women's long hair, whitening them, aging them. It buried anything that did not keep moving. They began to lose sensation in their toes and fingers and faces. They knew enough of extreme conditions to know that frostbite would soon begin to burn and blacken their extremities if they did not take action.

Then William Eddy remembered a bit of frontier lore he had picked up from some Rocky Mountain trappers. He gathered together the twelve people still alive and instructed eleven of them to lie in a circle on the snow, as closely packed together as they could get, with their feet pointing in toward the center of the circle. The twelfth person was to sit upright in the middle of the circle, at the intersection of all the feet. Eddy then arranged their blankets over their heads so that the person in the middle of the circle held one end of each blanket aloft. The other end of each was draped over the head of its owner and secured there with bits of wood or with snow. The end result was a low, circular tent that would entrap their body heat and shelter them from the biting winds.

It worked well. The snow falling on the outside of the blankets soon added its own insulating value to the shelter. Maintaining it required no more expenditure of energy than to change the person in the center from time to time and occasionally shake excess snow off the blankets. They began at least to feel sensation coming back into their extremities.

Outside, the storm moaned and whistled through the pine trees all night. The sun rose at 7:18 on Christmas morning, but the storm continued to rage, and as the day wore on, it showed no signs of abating. Psyches that had begun to crumble outside the shelter continued to do so within it. Breathing one another's exhalations, lying in the stench of one another's filth, cramped by starvation, they listened wearily as Patrick Dolan in particular continued to mutter and to rave. When they tried to sleep, they dreamed of food. When they awoke, they heard Patrick Dolan still shrieking.

Dolan thrashed about under the blankets. He began stripping off his clothes. Then he tried to crawl out into the storm. Eddy struggled to wrestle him back under the blankets, but Dolan wriggled free, crawled out of the shelter, and floundered off into the blowing snow, half naked. Eventually he returned to the shelter but simply lay down in the deep snow outside until the men dragged him back in and held him down.

Survival psychologists call reactions like Patrick Dolan's attempt to run away from the others the "hide-and-die syndrome" or "terminal burrowing." Like the paradoxical undressing that Dolan also exhibited, it is indicative of the final—terminal—stage of hypothermia, and between 25 and 50 percent of hypothermia victims experience it before they die.

By late afternoon Dolan's breathing grew shallow, his body grew rigid, and he died. The men dragged his body out into the snow and laid it alongside those of Franklin Graves and Antonio.

At the lake camp, Christmas Day brought little cheer, little to celebrate.

Just how much each of the families huddled there felt the bitter irony of spending Christmas in such miserable circumstances depended to some extent on where they and their forebears were from. Christmas as we know it was, in some senses, just being invented in America in the 1840s. Families that hailed originally from New England, as did Sarah's family, might still have felt some sense of the strong disapproval with which their Yankee parents and grandparents and their Puritan ancestors had regarded any special treatment of the day. But by the end of the 1830s, the old severe views were beginning to slowly give way, even in New England. For the first time, meetinghouses were beginning to be decorated and ministers were beginning to preach sermons on Christmas themes. The thawing would take a long time, though. As late as 1869, schoolchildren in Boston could still be expelled for skipping school on Christmas Day.

But German and Irish immigrants were bringing different attitudes to the United States in the 1840s, slowly altering the American concept of Christmas and gradually popularizing not just the Christmas tree—which until then had been found only in German settlements in Pennsylvania—but also the traditions of gift giving, feasting, decorating homes, and celebrating Christmas services with something of the elaborateness and joyousness of the Catholic liturgy. In 1842 the first commercial Christmas cards were printed. When Charles Dickens, immensely popular in America, published *A Christmas Carol* in 1843, it infused the American imagination with the revival of English traditions then under way in Victorian England, traditions like celebrating the day with roasted fowl, plum puddings, and the singing of carols.

By the mid-1840s, well-to-do families, particularly in New York and in the South, had begun to observe Christmas on a fairly elaborate scale. Arlington House, Robert E. Lee's home overlooking the Potomac in Virginia, was decorated with holly, ivy, and mistletoe. Gifts were given to family members and the household staff. A Yule log was set ablaze in one of the fireplaces, and the family attended special Christmas services at a nearby Episcopal church. On that particular Christmas Day in 1846, as the Donner Party huddled in the Sierra Nevada, Lee himself was fighting the Mexicans and living in a tent in Mexico. But a letter he wrote home to his family that day gives a sense of what Christmas at the Lee household was like.

I hope good Santa Claus will fill Rob's stockings tonight, that Mildred's, Agnes's, and [Annie's] may break down with good things. I do not know what

he may have for you and Mary, but if he only leaves
for you one half of what I wish, you will want for
nothing.

Later that day, even in a tent in Mexico, Lee sat
down to a feast of roast turkey and chickens and eggnog
at a table decorated with oranges and pine boughs.

For most Americans of a mind to celebrate the holiday that year, though—in towns and villages like those
that the Reeds and Graveses had come from—the
Christmas observations had a more homespun flavor.
Women baked cakes and other treats and slipped them
into their children's caps and stockings that night. Men
gathered and drank whiskey and hard cider and brandy.
Sometimes they fired off guns or set off firecrackers
and exploded water-filled hogs' bladders as they did on
the Glorious Fourth. Children attended socials or performed plays. Family members exchanged gifts—often
homemade treasures like quilts that they had labored
on throughout the year. Many of them sat down to a
turkey dinner, lending an American twist to the English tradition of roast goose. And almost always they
attended church, for in the end Christmas was above
all a religious occasion for most of them, an occasion on
which to contemplate the light that their faith brought
to the darkest time of the year.

. . .

For the Donner Party, it was exceedingly hard to find the light that Christmas, though. At the lake camp, Patrick Breen was too enfeebled by hunger and too incapacitated by kidney stones even to gather firewood. Devoutly Catholic, he and his family strove to maintain their faith. In their cold, dark shanty, Patrick sat down to his journal and wrote about his family's observance of the day, that they "offerd our prayers to God this Cherismass morning the prospect is appalling but hope in God Amen."

One of the most appalling prospects that faced Patrick Breen that morning was what seemed to be the imminent death of Augustus Spitzer, who still lay prostrate, barely clinging to life in a corner of the cabin.

In the Graves-Reed double cabin that morning, Margret Reed served her children the same gluey concoction of boiled ox hides that they had largely been subsisting on for weeks now. But later in the day, she had a holiday surprise. She had hidden away a few dried apples, some beans, a bit of bacon, and some tripe from the slaughtered oxen. These she slowly and carefully prepared and then laid before her wide-eyed children for their Christmas dinner. "Children," she cautioned, "eat slowly, . . . for this one day you can have all you wish."

On the other side of the log partition dividing the cabin, an almost identical scene played out. Franklin Graves had brought along a sack of dried beans with which he planned to begin farming in California. Elizabeth Graves had hoarded away a meager ration of them. And like Margret Reed she had also kept a small amount of tripe buried in the snow to simmer with the beans in her large Dutch oven and lay before her children for Christmas dinner.*

Despite their miserable circumstances, Elizabeth Graves had good reason to cling to hope that Christmas afternoon. She still had a stock of frozen beef buried in the snow. Franklin and the girls had been gone for nine days now on a journey they expected to take no more than six days. For the first few of those days, there had been clear skies and a firm crust of snow to walk upon. She could reasonably assume that by now he was at Johnson's Ranch laying in provisions, or perhaps already starting back over the mountains, bringing those provisions to her and her children.

In the Murphy cabin, there was less reason for hope. Out of beef, with their supply of hides rapidly

---

* It is possible that Margret Reed and Elizabeth Graves collaborated on Christmas Day to share the treasures they had hoarded. The fact that both managed to produce tripe for dinner suggests that they might have overcome the tensions that were building between them to make the day special for their children.

dwindling, Levinah Murphy and Eleanor Eddy had taken to gathering the cast-off bones of slaughtered oxen. They boiled the bones and served the broth to the nine children huddled in the cabin. Then they boiled the bones again and again, until they became soft enough to chew, and served them whole to the children. On this Christmas Day, Levinah Murphy added a few pieces of oxtail to the broth, to make it perhaps a bit more festive.

They ate bones for Christmas at the Donners' camp at Alder Creek as well, as they had been doing for a while now. Sometimes they boiled them; sometimes they toasted them brown in the coals and then gnawed on them. And as at the lake camp, they boiled hides. Insofar as they had anything to share, Tamzene and George Donner shared with the now-widowed Elizabeth Donner and her children and with Doris Wolfinger. But there was precious little to share. In their insubstantial shelters, they all lived in almost perpetual danger of hypothermia, sometimes lying abed for days at a time in wet clothes, trying to keep smoky fires going through the relentless cycles of freezing rain and snow, constantly brushing snow from their tents lest they be buried. Most of the healthy young men on whom they had depended were now gone, either dead and buried in the snow or departed for the lake camp. George Donner, with the infection from his cut hand still climbing relentlessly up his arm, could do almost nothing to help his wife and children.

On Christmas night the storm eased, and it was over by the morning of December 26. On the slope leading down into the canyon of the North Fork of the American River, Sarah and Jay and the others peered out from under their blankets that morning and found themselves surrounded by deep drifts of fresh snow. Without the hatchet they had lost during the storm, they were unable to cut wood or start a fire, so they mostly stayed in their makeshift blanket tent, conserving their body heat and their energy during the morning. But in the afternoon they crawled stiffly and painfully out of their shelter. For some time William Foster was so stiff that he could not get his limbs to unbend at all.

They spent much of the remainder of the day hunting for the hatchet, their snowshoes, and their packs, all of which had disappeared under the snow. As they tried to gather wood without the hatchet, someone broke a dried pine branch from a tree, and a mouse ran out and scurried away. They all chased it, shouting and thrashing through the soft snow in pursuit of it. Thirteen-year-old Lemuel Murphy, who was growing increasingly demented, seized the mouse, thrust it into his mouth, and ate it alive.

The three bodies lying nearby—Franklin Graves, Patrick Dolan, and Antonio—were rigid and blue and half covered by snow. The living avoided them. They already knew what they were going to do, but they were not yet ready to do it.

To some extent they had become apathetic. The worst hunger pangs had begun to pass after thirty-six hours without food, and while they were in many ways physically miserable, hunger was not always the most pressing component of their misery. Their brains had stopped screaming out for glucose, partly because their bodies had made some critical adjustments, designed to preserve the integrity of their brains for as long as possible. Their guts had begun to shrink, reducing the surface area of their digestive systems. Their livers had begun to transform fatty acids into chemical compounds called ketone bodies. These were able to mimic glucose and provide their brains with up to two-thirds of the energy they needed to function. The use of ketone bodies carried a price, though. It was gradually acidifying their blood, leading them toward a dangerous condition called ketoacidosis, common in diabetics. The most obvious manifestation of the condition was that large amounts of acetone were being released in their urine and their respiratory systems. As they huddled under their blankets, their breath began to smell like something they had never known—fingernail polish.

Late in the day, they crawled back under the blankets, still unfed except for Lemuel. The pittance of nourishment provided by the mouse seemed to stimulate the boy's madness and renew his hunger pains, though. As night fell, he howled and raved and grabbed at people's arms, biting them, crying out, "Give me my bone!" His sister, twenty-year-old Sarah Foster, held him tight and tried to comfort him, but, like Patrick Dolan the night before, he clawed his way free and scrabbled about, bent on escaping from the tent. Finally they all forced him to the center of the circle with their feet, trying to keep him from slipping out under the perimeter.

The skies were clear that night, and a waxing gibbous moon crossed above the rim of the canyon. Everything in the canyon—still and white and crystalline—shimmered. Under the blankets, Lemuel Murphy finally quieted down. His sister, sobbing, held his head in her lap until, at about 2:00 A.M., he ceased breathing. Then they rolled his body out into the moonlit snow and closed the circle tighter, down to ten now.

The next day they set about the task of butchering meat.

THE HEART ON THE MOUNTAIN

The first order of business on the morning of December 27 was to make a fire. Under the blankets, William Eddy poured some black powder from his powder horn onto a bit of tinder and at the same time struck a spark from his flintlock rifle. Pouring black powder onto tinder, especially if the tinder is damp, is an old woodsman's trick for increasing the likelihood that the spark will catch. Unfortunately for Eddy, it caught with a vengeance, exploding the powder horn in his hands with a terrific flash of smoke and flame. He scrambled out from under the blankets with a blackened face and badly burned hands. Amanda McCutchen and Sarah Foster followed him out, also burned but not so badly. Eventually they got a fire kindled in dried branches they had collected and used it to set fire to another dry dead pine. Then they began doing what they had by now agreed that they would do.

They divided into groups so that no one would have to eat, or see eaten, any of their kin. Sarah and Mary Ann and Jay stayed apart from Franklin's body, Sarah Foster from Lemuel's body. Luis and Salvador would have no part of any of it. They built a separate fire at a distance and turned their backs on the whites.

If they did as others in similar circumstances have almost universally done, Jay Fosdick, William Eddy, and William Foster started by removing and concealing in the snow the heads, hands, and feet of the dead, to render the bodies a bit less human. Then, as they would with a deer or an ox, they cut open the body cavities and extracted the most nutritious organs: the liver, the heart, and the kidneys. These would not keep well; they needed to be eaten first.

Now that they had crossed the line, their hunger put itself foremost in their thoughts. So at some point shortly after they had taken these organs from the bodies, they stopped and sat down to their first unthinkable meal. They put the meat on sharpened stakes and held it out over glowing coals, roasting it until they judged it done, or done enough. The smell of roasting meat is largely the same no matter what type of meat, and, unbidden, it stimulates the appetite mightily, activating the salivary glands, awakening the gut, grabbing the attention of the brain. So when it had cooled enough that it did not burn their lips, they sat down in the snow, weeping, their eyes averted from one another's faces, and took their first few tentative bites. Then they ate.

And when they ate, their digestive systems gurgled and surged back to life and demanded more, and so they ate more. Their headaches and bone-crushing weariness began to lift. Energy poured into their limbs. So they got up and prepared more of the flesh and ate more, still avoiding one another's eyes as best they could. For the first time in days, they now believed they would live at least a few more days, but they also knew that for the rest of their lives they would bear a terrible awareness of what they had done here on this day. For three of them—Sarah, Mary Ann, and Sarah Foster—the psychic burden was all the more crushing for knowing that at one of the adjoining campfires someone was at that moment eating their father or brother.

Most people faced with starvation, most of the time, choose to die rather than resort to cannibalism. The prohibition against eating human flesh is as ancient and fundamental a taboo as can be found. That is not to say that cannibalism is rare in the history of the world, though. It has been practiced, sometimes on a very large scale, in nearly every corner of the world and nearly every age. Neanderthals are believed to have chowed down on one another from time to time, and early *Homo sapiens* likely did as well. There are biblical accounts of cannibalism, tenth-century accounts of Christian Crusaders eating captured Arabs, and widespread accounts of cannibalism among indigenous peoples of South America, Polynesia, and North America. Much of this anthropophagy, to use the technical and more euphemistic term, has had nothing to do with survival, but rather with ritual and religion. Or sometimes simply with vengeance.

But there have also been large-scale examples of survival cannibalism, many of them in disturbingly recent times. During the great famine in the Ukraine in the early 1920s—a horrific catastrophe that caused some 5 to 8 million deaths—so many dead bodies, particularly the bodies of children, disappeared from the streets that authorities had to put up signs proclaiming EATING DEAD CHILDREN IS BARBARISM. During the 900-day siege of Leningrad in 1941–44, people resorted to eating first dogs and cats and then finally rats. When the rats were gone and human bodies began piling up in the streets, many of them were soon stripped of their flesh. As things got even worse, the Leningrad police had to track down organized gangs who had gone into business kidnapping, murdering, and butchering their victims for meat. Once again it was children who disappeared fastest. In the apartment of one violinist, authorities found the bones of several dozen children. The violinist's own five-year-old son was among his apparent victims.

Even the Russian and Ukrainian catastrophes paled in comparison, though, with the appalling horror that descended on the Chinese people between 1958 and 1962. A combination of drought, floods, and the economic policies of Mao's Great Leap Forward caused some 30 to 40 million of them to starve to death. By as early as 1959, the famine was so widespread in some rural parts of China that peasants began to eat the corpses of their fellow villagers, particularly the corpses of children. When they ran out of corpses, some families took to starving their infant daughters and then exchanging the bodies with those of their neighbors' daughters so that nobody would have to eat his or her own children. They made soup out of them.

Still, the fact is that cannibalism is a remedy that remains well beyond the last resort for most of us. For every poor soul who has eaten of a companion, there are countless who preferred to die so that a loved one might live.

So the question arises: Why did the men of the snowshoe party draw out their knives on the morning of December 27, 1846, and commence carving? And why did Sarah and the other women eat what the men carved from the bodies? They had been entirely without food for just six days at most. People have lived far longer than that without food, even in very cold environments. Despite the tremendous rate at which they were burning calories, they likely had considerable time to go before they actually starved, perhaps weeks.

It may have been, for one thing, that they were at least partly mistaken as to what was killing them. Their growing malnutrition was rapidly breaking down both their psychological defenses against madness and their physiological defenses against the cold, all of which must have contributed to an overwhelming sense of impending doom. To some extent they believed that Stanton, Antonio, Franklin Graves, Patrick Dolan, and Lemuel Murphy had died of hunger, when in fact they almost certainly had died primarily of hypothermia. As a consequence the survivors might well have thought that they themselves were starving to death—it certainly must have felt as if they were. Then, too, they had already broken through the psychological barrier that ordinarily prevents us from seeing food when we look at one another. When Patrick Dolan had drawn the fatal lot, they had all been given license for the first time to turn their hungry eyes on a fellow human being and see a potential meal.

And in the seemingly endless hours that the snow-shoe party had spent under the blankets, we know that some of them had begun to have what clinicians call anthropophagic dreams—visions of eating other people—a phenomenon not uncommon among those facing starvation.*

Mostly, though, what allowed the men of the snow-shoe party to pull out the knives and what allowed both the men and the women to eat was likely something beyond an instinct for self-preservation and something well short of madness. All of them—except possibly Luis and Salvador, about them we just don't know—had someone back at the lake camp whom they loved, someone who was depending on them to get through and send help. Their own survival meant more to them than simply continuing to live. It meant hope for those they had left behind. And Sarah and Mary Ann must have felt this obligation particularly keenly. Only a few days before, their father had laid a sacred charge on them to save the lives of their mother and siblings, instructing them explicitly to do whatever was neces-

---

* At least one of the young men enrolled in the Minnesota hunger experiment—Franklin Watkins—suffered vivid dreams of eating insane people, just before he himself descended into hunger-induced psychosis.

sary. And nobody had to tell Harriet Pike and Sarah Foster and Amanda McCutchen that the lives of their infant children back at the lake camp depended on their making it to Johnson's Ranch.

None of that made it any easier, though.

When they finished eating and finally began to look at one another again, and to talk, they started to make plans. The weather had cleared and grown colder, and that meant that there was a crust on the snow. They had no way to know how long the favorable weather would hold, but they knew that they should resume their trek as soon as possible. First, though, they had to render the remaining flesh portable and nonperishable. They cut off long, thin strips and stretched them on racks or stakes placed before the fire. It was a long, slow, gruesome process. They had to take care to keep the flesh close enough to the fire to dry it but not so close as to actually cook it. So for three days, under mostly clear skies, they worked at the task, and rested, and continued to eat.

By the morning of December 29, they were ready to travel again. They loaded their packs with their blankets and what remained of their former companions, strapped on their snowshoes, and struck off again. The four bodies had yielded less meat than they might have expected. Like the flesh of the oxen they had all been subsisting on since November, the flesh of the four malnourished corpses was lean and stringy. It contained almost no fat. A man as large as Franklin Graves might have yielded as much as sixty-six pounds of fresh, usable meat when healthy, but his body had probably yielded roughly half that by the time he died. An emaciated boy like Lemuel Murphy might have yielded as little as twenty pounds. Once the flesh had been dried, of course, it weighed a fraction of these amounts. And they had been living on it for three days already. By the time they left what would come to be called the "Camp of Death," they calculated that they had only about four more days' worth of the grisly rations.

The weather remained good for traveling, clear and cold, and with their bodies now resupplied with fuel, they made about five miles that day, and six the next. Seeking to hold to a roughly southwesterly course and navigating by the sun, which was at this time of year well to the south of due west, they inevitably worked their way deeper into the drainage of the American River's North Fork. By December 31, though, they found themselves boxed in by high cliffs ahead and the increasingly deep main canyon to their left. They decided to cross the gorge itself and proceed southwest along the ridge visible on the other side.

The descent was steep. It was so steep, in fact, that soon they could not keep their balance when walking. They found, though, that by squatting on their snowshoes they could simply ride them downhill like sleds. This saved enormous amounts of energy, but it also caused them, time and again, to go out of control, tumbling head over heels into heavy drifts of snow at the end of each run, driving snow deep inside their increasingly tattered clothing, next to their bare skin.

By the end of the day, they had made it down to the river, and on the morning of New Year's Day, they set out to ascend the steep—in places almost vertical—opposite side of the canyon. It took them an entire day of laborious, leg-throbbing climbing. With each step they had to drive the toes of their snowshoes into the wall of snow before them, then lift themselves up the slope as if they were on a ladder, much as a modern mountain climber does, but without the benefit of crampons. Where the climb was steepest and there was little snow clinging to the rocks, they scrabbled desperately for footholds in the clumsy snowshoes, grabbing at roots and branches to keep from plummeting back down into the canyon.

By the time they reached the top that afternoon, their feet were swollen and cracked and bleeding. Their leather shoes were so rotten now that they were falling to pieces. They wrapped their feet in rags and bits of blanket, but the blood oozed out through the fabric and into the snow, leaving crimson tracks wherever they went. The frostbitten toes of one of the Miwoks had begun to fall off at the first joint.

Exhausted by the ordeal, and famished, they ate again, digging deeper into their packs for the dried flesh. From the high ridge on which they now found themselves, they saw glimpses of what they took to be the broad Sacramento Valley, still far off to the west. But they also saw ridge after snowy ridge stretched out between it and them.

The climb had taken much out of all of them, but the men in particular were fatigued and dispirited. Since leaving the Camp of Death, the women had more and more often found themselves having to take the initiative, setting the course, sometimes leading the men by the hands, gathering the wood, making the fires at night, bringing them food.

A Donner Party survivor later told J. Quinn Thornton, author of one account of the snowshoe expedition, that the men had been ready to give up well before the women.

The deep stupor into which their calamities had plunged the most of them often changed to despair.

Each seemed to see inevitable destruction, and expressed in moans, sighs, and tears the gloomy thoughts over which their minds were brooding.

Of the women, though, the same survivor said,

Most of them manifested a constancy and courage; a coolness, presence of mind, and patience. . . . The difficulties, dangers, and misfortunes which seemed frequently to prostrate the men, called forth the energies of the gentler sex and gave them a sublime elevation of character, which allowed them to abide the most withering blasts of adversity with unshaken firmness.

The women, of course, were themselves in mortal danger hour by hour, but if the men were not going to break down, it seemed that it was the women who would have to make sure of it. By the time they arrived at the top of the high ridge late on the afternoon of January 1, Sarah must have been watching Jay closely, studying him for clues.

As they began to push forward on January 2, both the men and the women found that they could make better progress without the snowshoes, so they took them off and strapped them on their backs. They hobbled forward, but their feet were now so swollen that every step was excruciating and their progress was slow. The country was gradually changing as they moved southwest and downslope. Granite peaks gave way to rounded mountains. A mix of oaks and long-needled digger pines began to replace the tall conifers of the high Sierra. The snow cover grew lighter, and here and there bare patches of red, gravelly earth began to show through beneath.

When they made camp that evening, they were near the end of their rations again. Some of them toasted the remnants of their leather shoes over the coals of the fire; others disassembled their snowshoes and toasted and ate the rawhide strings.

On January 3, gradually losing elevation, they began to encounter country that was largely snow-free for the first time, but in place of the snow they found a new impediment. The pale, silvery green leaves and red, twisting branches of head-high manzanita brush covered the steep hillsides here. As they fought their way through dense stands of the manzanita, the branches caught on their already half-rotted and tattered clothes and began to rip them apart.

By January 4, Jay Fosdick had started to lag far behind the others. Sarah fell back to stay in step with him.

Time and again the others found themselves having to stop and wait for the two of them. William Eddy studied Jay's halting progress and finally went to his side and told him flat out that he was going to die if he did not exert himself more. They had been lucky enough to have fair skies for over a week now, but nobody could tell how long their luck would hold.

They pushed on, watching all the while for any sign of humanity, whether white or Native American. They had little to fear from the local Indians, the Maidu. Intimidated by what they had seen the Mexicans and John Sutter do to their Miwok neighbors in the lowlands, the Maidu mostly just wanted to live in peace in these hills that, so far, none of the whites had found any reason to covet.

By that evening the snowshoe party had had no food for several days, except for bits of toasted leather, and the mood among them began to grow ugly. William Foster first brought up an idea that had likely festered in the minds of at least some of them for days. Why not kill Luis and Salvador for food? While the Miwok boys were at least technically Christians, they remained in the eyes of some of the whites, if not all of them, savages nonetheless—the same general class of beings many had come to loathe during the Black Hawk War of their youths. Looked at a certain way, they were ignorant, itinerant beggars at best, dangerous cutthroats at worst. Looked at another way, they were simply strangers. When killing to survive, it's easiest to kill and eat whatever or whomever you are least attached to—cattle before horses, dogs before people, strangers before acquaintances, acquaintances before friends, friends before family. Luis and Salvador, more than any of the others, were strangers to them all.

They mulled it over, discussing it in low tones, watching the Miwok boys out of the corners of their eyes. Not everyone agreed with the plan. William Eddy argued against it. What Sarah said or thought, we do not know.

Finally Foster abandoned the idea, at least for the time being. Eddy said that in the morning he would go ahead with the gun and look for game. Now that they were below the snow line, there was a reasonable chance that he might be able to kill a deer. Later that night Luis and Salvador slipped quietly away into the darkness and disappeared. It may be that Eddy warned them, or they might simply have noted the darkening looks in the haggard faces of the whites.

The next morning, January 5, limping through the chaparral, Mary Ann Graves and William Eddy went out ahead of the others, carrying the flintlock rifle, looking for game. Here and there among the manzanita, they could make out Luis's and Salvador's bloody footsteps. Harriet Pike, Amanda McCutchen, and Sarah and William Foster followed in a second group. Sarah and Jay Fosdick brought up the rear, once again quickly falling behind the others.

A mile or two out of camp, Eddy and Mary Ann Graves came across a place where a deer had recently lain in the brush. The sudden discovery of exactly what they had hoped for stopped them in their tracks, stunning them with an unexpected mixture of desperate hope and profound dread. They glanced at each other, and each discovered that the other had begun to weep. They dropped to their knees and prayed, then rose and staggered on as quietly as they could in the brush, stalking the animal.

They followed the deer's tracks until they spotted it browsing about eighty yards off. Eddy moved closer, angling for a clear shot. He raised the rifle, but his arms were too weak to hold it level and straight. He lowered it and raised it again and heard Mary Ann Graves behind him give out a little sob. He turned and looked at her and saw that she had covered her face with her hands.

When you fire a flintlock rifle, a small but disconcerting delay follows between the time you pull the trigger and the time the shot goes off. The cock holding the flint must fall and strike the steel of the striker plate. Then the resulting spark must fall into the pan and ignite the powder. The flash in the pan must then penetrate the touchhole and ignite the powder in the chamber, producing a second, larger explosion. This second explosion must then finally propel the ball out of the barrel. All told, it might take nearly a second to get the shot off, a small eternity when you are trying to hold your aim steady and true on a distant target. And a good many times, if the powder is damp or a breeze blows the spark away, nothing happens at all when you pull the trigger.

Eddy raised the rifle yet again, aiming the muzzle well above the deer this time, then letting it slowly descend until the deer fell into his sights. He pulled the trigger, and the gun discharged. The deer leaped a yard or two and stood still, dropping its tail between its legs. Mary Ann cried out, "Oh, merciful God, you have missed it!" The deer bounded forward. Eddy dropped the gun, and he and Mary Ann took off in pursuit, limping as they ran through the brush. Two hundred yards later, the deer tumbled to the ground, dying. When they reached it, Eddy drew out a knife and slit its throat. He and Mary Ann knelt in the chaparral and drank the blood as it spurted out of the animal's veins.

. . .

Far behind them, Amanda McCutchen, Harriet Pike, and the Fosters heard the shot that killed the deer and began to hurry forward. Even farther back, Sarah and Jay also heard it. But they had come to a standstill perhaps a mile behind the second group. When Jay heard the shot, he cried out to Sarah, "There! Eddy has killed a deer. Now if I only can get to him, I shall live." But he could barely stand by now, let alone walk any distance.

That night Mary Ann Graves and William Eddy feasted on the roasted entrails of the deer. During the night Eddy fired the flintlock at intervals to let the others know where he and Mary Ann were camped, trying to guide them in. But Amanda McCutchen, Harriet Pike, and the Fosters did not make it to the deer that night. They camped on a ridge above them and endured another long night of hunger cramps.

Farther to the east, Sarah wrapped Jay in the one blanket they owned and sat down beside him where he lay on the ground. Now and then she heard the report of Eddy's rifle off to the west. She might have struggled on alone to Eddy's camp, and food, guided by the sound of his gun, but she could not bring herself to leave her husband, who seemed to be slipping away.

Jay likely did not hear the later rifle shots, nor Sarah speaking to him at his side. By now he was suffering severe malnutrition and probably severe hypothermia as well. When he spoke to her, his speech was slurred. It was another night of bright moonlight, and his skin must have been pearly white and cool under her touch. His face was cadaverous and shrunken, his frame gaunt and disjointed beneath his clothes. As the evening wore on, his breathing grew shallow and rattling. His mouth and lips grew dry. His heartbeat grew erratic. Time moved ever more slowly for him, and for Sarah sitting beside him. He lapsed into and out of consciousness. Finally, a little before midnight, he died. When Sarah was sure he was gone, she rolled his body in the blanket and then lay down beside him, and held him, and tried to die herself.

But Sarah's heart continued to beat and she to live. In the morning she took a black silk neckerchief from around Jay's stiff neck and put it around her own and then staggered forward, alone now, toward California.

Before long she came across William and Sarah Foster and Mary Ann, who were backtracking, looking for her and Jay. When Sarah told them that Jay was dead, the Fosters wasted no time. They asked Sarah point-blank if they might eat him. Sarah must, by now, have been beyond any expectation of sympathy from her companions. She must in fact have been beyond any expectation of any sort of mercy from the indifferent Fates. She looked at the Fosters and said simply, "You cannot hurt him now," and continued up the trail with Mary Ann. The Fosters went on to where Jay's body lay and began to butcher it, severing his legs and arms from his trunk, packing onto their backs as many pieces of him as they could carry.

When Sarah reached Eddy and the remains of the deer, she ate roasted venison, the first food other than human flesh that she had tasted since December 21. Harriet Pike and Amanda McCutchen caught up with the others and joined in eating the venison. Then the Fosters also came into camp, bearing their grisly burden.

That evening they all sat around the campfire in a small circle of wavering light. Eddy had already begun drying strips of the venison by the fire, but the meat that the Fosters had brought into camp was fresh, and apparently too tempting to resist. Someone sharpened a stake, impaled Jay Fosdick's heart on it, and held it out over the coals.

What Sarah thought and felt we can only try to imagine, if such a thing is even possible. She never wrote or spoke about it in so far as we know. She might have hidden her eyes; she might have left the camp-fire. But there was not very far that she could go in the dark tangle of manzanita around her, and the aroma of roasting meat must have filled the night air far beyond the circle of light. Wherever she took refuge, the vast, silent, suffocating blackness of the California foothills began to close in on Sarah, more alone now than she had ever imagined she could be.

11

MADNESS

As Sarah lay in the cold darkness of the California foothills that January night, living the nightmare in which she found herself, the larger world of course went on without her.

Like all tragedies, hers took place in a historical context, and as is often the case the context sheds light on how Sarah must have viewed her own situation as it unfolded. In many ways she and Jay had been moving backward in time as they moved westward across the continent. They had been slowly leaving behind the modern world as it was then and walking into the essentially Stone Age world in which the California Indians had lived for millennia. Sarah had always lived on the frontier, but even in Illinois her world had been connected to the larger world of American commerce and ideas. Here on the western flank of the Sierra Nevada, though, she found herself in a world devoid of virtually all the creature comforts and technologies of her time.

The mid-1840s was not so divorced from our own time as we sometimes tend to think when we peer at the often haggard faces of emigrants in dusty old daguerreotypes. In significant respects, in fact, the period marked a transition between a much older way of life and a new era of innovation in which we still live. The Industrial Revolution was already well under way in Europe and in the urban centers of the eastern United States, and a new spirit of scientific inquisitiveness was being applied to nearly every aspect of life. Just two years before Sarah had set out for California, Samuel Morse had successfully transmitted a message—"What hath God wrought?"—across a thirty-eight-mile telegraph line from Baltimore to Washington, D.C. In September of 1846, as Sarah made her way across the sage-lands of Nevada, *Scientific American* had published its first issue. That same month Neptune was discovered—not by random searching of the sky but by the use of mathematical models to predict its location. The following month, as Sarah first entered the Sierra Nevada, Dr. John Collins Warren made the first public demonstration of an effective anesthetic—ether—to painlessly extract a tumor from a patient's jaw. The American inventor Richard Hoe developed a rotary printing press that year that could spit out eight thousand printed pages per hour. In England, Daniel Gooch unveiled a powerful new steam locomotive—the Great Britain—that could pull a staggering one hundred tons of deadweight at fifty miles per hour for seventy miles.

And it wasn't just the technological foundations of our world that were being laid in the 1840s; cultural issues and concepts were arising that would dominate the twentieth century and live into the twenty-first. Charles Darwin had published *Voyage of H.M.S. Beagle*. Karl Marx and Friedrich Engels were at work on *The Communist Manifesto* even as Sarah suffered in the mountains. On the East Coast, women's benevolent societies were planting the seeds of modern feminism. Richard Wagner was inventing new concepts in the language of music, composing operas like *Rienzi* and *Tannhäuser*. Edgar Allan Poe was writing the first modern detective story, "The Murders in the Rue Morgue," and the first modern mystery, "The Gold Bug."

If Sarah had been picked up out of the foothills of the Sierra Nevada, flown twenty-five hundred miles east, and put down in New York City, she would have found a scene of enormous vitality and considerable modernity. She would have moved among throngs of people bustling through a network of busy streets—streets that you or I could follow using any twenty-first-century map of lower Manhattan. On the southeast side of town, she would have found herself among new, monumental stone buildings, many of which still stand today. At 26 Wall Street, she could have visited the new Customs House, now the Federal Hall National Monument. At 55 Wall Street, she would have found businessmen hurrying in and out of the stately new Greek Revival Merchants' Exchange Building. Up the street and around the corner, she could have stopped to offer prayers for Jay's soul at Trinity Church, the soaring Gothic Revival spires and steeple of which made it the tallest building in the city. If she had climbed into the steeple, she could have looked out over sprawling shipyards and watched dozens of steam ferries plying the East River between Manhattan and Brooklyn, watched passenger liners belching clouds of black smoke as they departed for Europe. She could have taken a commuter train from City Hall north to Twenty-seventh Street for six and a half cents or continued to Harlem for twelve and a half cents. She could have taken the Long Island Rail Road north from Brooklyn all the way to Boston via a train/ferry combination. She could have watched students poring over their books at Columbia College on Park Place or at New York University, or sat on adjoining Washington Square watching gaslights flickering all around her as dusk gave way to darkness.

Much more likely, though, Sarah would have headed straight for one of New York's 123 full-service restaurants, perhaps an economical choice like Sweeney's on Anne Street, where she could buy a plate of assorted meats for six cents, a plate of vegetables for three cents. If she had one of the silver dollars from her father's hoard and wanted a fine beefsteak, she might have stopped in at Delmonico's on Beaver Street. Even if she had no money at all, she could have retreated to the Alms-House, a sprawling thirty-acre complex on the East River, complete with sixty sleeping apartments, two large dining rooms, a school, a chapel, and a hospital, open to any "well-behaved person" who might apply for aid and a hot meal.

But there were no almshouses in the foothills of the Sierra Nevada. There were no telegraph wires to send an SOS to Johnson's Ranch; no steam train to carry Sarah to Sutter's Fort; no restaurants of any kind, let alone Delmonico's; no flickering gaslights, only the flickering of the campfire over which Jay Fosdick's entrails had just been roasted. The most advanced technology at the disposal of Sarah and her companions was the old flintlock rifle that Eddy carried. They were

no better equipped, technologically, than Sarah's forefathers had been when they first stepped onto the continent two hundred years before. If any of them were
going to survive the rest of the journey, it would have
to be on the basis of strength, endurance, cunning, and
courage.

On January 7, Sarah picked up what remained of her
hope and her life and staggered on, following the others
forlornly through the manzanita. They were down to
seven now, all five of the women but only two men—
William Eddy and William Foster. Before they had
gone far, they discovered that the river, deep in the
canyon below them, had made an abrupt ninety-degree
turn to the south. To continue on a generally westerly
course, as they knew they had to in order to have any
chance of finding Johnson's Ranch, they would have
to recross the canyon. Over the next many hours, they
slid, stumbled, and fell into the gorge, descending from
an elevation of roughly twenty-seven hundred feet to
the river at about twelve hundred feet.

To a large extent, they were barefoot now, their
shoes having disintegrated and the shreds of blanket in
which they had since wrapped their feet rapidly falling
apart as well. By the time they reached the bottom of
the canyon, their feet—already cracked and swollen—

had been lacerated by the sharp rocks of the canyon wall. Their clothes had been burned through in places from crowding too close to their nightly fires. Their garments were so tattered that neither the women nor the men could even begin to maintain their modesty any longer—thighs and breasts and buttocks peeked out from under the miserable rags that hung limply from their shoulders.

They camped that night in country that was laden with gold. Hundreds of millions of dollars' worth of gold lay in the California foothills, much of it right where they now slept—hidden in the blue gravel beneath them, in pockets behind boulders deep in the fast-flowing river, in the sharp quartz rocks that had been cutting their feet and their hands all day, in the ancient drifts of sand and gravel hundreds of feet thick, forming many of the ridges around them.* In a little more than a year, John Marshall and Peter Wimmer—the latter of whom had traveled just ahead of Sarah back on the plains—would pluck a gold nugget out of a mill-race at Coloma twenty miles south of here and transform California forever. Eighteen months from now, men would begin swarming up and down this river canyon and over these hills, first by the hundreds, then

---

* Roughly $7 billion worth of gold was extracted from the foothills of the Sierra Nevada during the Gold Rush, in 2009 dollars.

by the thousands. Sarah and her companions would not then have been able to travel a hundred yards up or down this river without coming to a campfire, a tent, a warm meal of bacon and beans and biscuits. But that was in the future, and they were not. For now this was as lonely and trackless a place as they could imagine, without comfort and without mercy.

The next day they faced the much harder task of climbing back out of the canyon on the northwestern side of the river. It was again an all-day effort, and this time it took nearly everything they had to make the fifteen-hundred-foot, nearly vertical climb. They grappled for footholds and handholds as brush lashed at their faces and further shredded their clothes. Their feet smeared the rocks with blood as they climbed. They concentrated on the rocks and brush in front of their faces, trying not to look back over their shoulders as they climbed higher and the void behind them grew increasingly alarming. They heaved and grunted, gasped and sometimes sobbed. Toward the end of the day, they crawled over the rim of the canyon and lay prostrate on relatively level ground. They were likely not much more than about twenty miles almost due east of Johnson's Ranch now, but they did not know that. They did not in fact know within a hundred miles where on earth they were.

. . .

That same day, at Truckee Lake, Margret and Virginia Reed staggered back into camp, along with Milt Elliott. The three of them had been five days undertaking an audacious but ultimately doomed attempt to break out of the mountains.

Before they had left, Margret Reed had taken the last hide from the roof of her half of the double cabin and moved her children to the Breen and Keseberg shanties. There they had made a meal of Cash, the last of the Reeds' family dogs, an experience Virginia later wrote her cousin Mary about.

> We had to kill little Cash the dog & eat him we ate his head and feet & hide & everything about him o my Dear Cousin you don't now what truble is. . . . There was 15 in the cabon we was in and half of us had to lay a bed all the time thare was 10 starved to death while we [were] there was hadley able to walk we lived on little Cash a week and after Mr. Breen would [cook] his meat and boil the bones Two and three times. . . .

Then they had set out for the pass. The Reeds' servant, Eliza Williams, had started out with them but became exhausted and returned the second day, but the others had stopped to fashion improvised snowshoes and made it over the pass. By the fourth day, though, they had become lost, wandering among the peaks. When Virginia's feet had begun to suffer from apparent frostbite, they had finally aborted the effort and returned, though Margret Reed well knew she had almost nothing more to offer her children in the way of food back at the lake.

In the cold darkness of the Murphy cabin huddled up against the large granite boulder, meanwhile, the widow Levinah Murphy had grown seriously ill and was often unable to arise from her bed. And she was beginning slowly to go blind—perhaps from snow blindness, perhaps from general debility brought on by foraging for food and wood. Her seventeen-year-old son, John Landrum, who as the oldest male in the family had for weeks been doing most of the heavy work, also lay in bed, starting to rant and rave. Her younger children were doing better, but all of the infants in the cabin—Catherine Pike, George Foster, and Margaret Eddy—were weakening rapidly. Catherine was especially feeble. Plagued not only by hunger but also by the lice and bedbugs that infested her bedding, she made low, barely audible sobbing sounds almost continuously now. As her face shrank in on itself, her eyes seemed to grow larger and darker.

Half a mile to the east, in her part of the double cabin, Elizabeth Graves tended to another rapidly declining infant. Harriet McCutchen was also afflicted by lice, and her screaming was so incessant that it haunted Patty Reed on the other side of the log partition day and night. Elizabeth's own children were still relatively robust, but her stock of beef was running low.

In the hills above the North Fork of the American River, insanity stalked the snowshoe party. They began to glare at one another out of hollow eyes, like wild animals. William Foster in particular seemed to be coming unhinged. Over the past few days, he had grown listless and despondent. He had ceased helping to make fires at night and looked more and more to his wife and the others to take care of him.

Precisely what happened next is unclear, as later accounts varied. By some accounts, after gazing upon Amanda McCutchen for some time, Foster approached William Eddy privately and proposed killing her for food, arguing that she was slowing them all down anyway. Eddy protested violently against such a thing, pointing out that Amanda was a mother. Foster, in this version of events, then reportedly proposed another plan that would avoid that objection and also provide a greater yield of meat as well—killing both Sarah and Mary Ann, neither of whom was a mother. Eddy then apparently pulled a knife and threatened to kill Foster if he pursued the subject. It's impossible to know whether the confrontation happened so dramatically or whether the story simply grew more vivid with the later telling. A different version of events suggested, in fact, that Eddy himself tried to lure Mary Ann Graves away from the others in order to kill her. But it is clear that Foster at least, and perhaps Eddy as well, had reached some kind of snapping point.

It's not surprising that minds had begun to unravel in the foothills of the Sierra Nevada. As Nathaniel Philbrick points out in *In the Heart of the Sea*, people living in conditions of extreme stress often undergo a process of psychic deadening. They stop experiencing ordinary human feelings such as compassion and understanding. Their desire to survive usurps these emotions and replaces them with a kind of cunning, cruel, self-centeredness. They begin to ignore the rules the rest of us play by. Religious and moral tenets that they may have adhered to all their lives begin to fall away, freeing them to do whatever seems in their best interest at any given moment.

They don't necessarily go through this process alone, though. Small groups of people living under these kinds of stresses and experiencing this kind of psychic numbness tend to form feral communities, tribes governed not by the usual social conventions but by older, more fundamental laws—the laws of necessity, dominance, self-preservation, and brute strength. Such communities evolved in German concentration camps during World War II, and they evolve in modern-day prisons. One of the hallmarks of feral communities is that they tend to splinter into subgroups based on distinctions of identity, such as gender, race, religion, and culture. This stress-induced rending of the social fabric often allows a stronger group to prey on a weaker one, higher-status individuals to exploit lower-status individuals. That is why black slaves were often the most at risk when shipwrecked with white crews in the eighteenth and nineteenth centuries, why Luis and Salvador had been wise to slip away from the snowshoe party, and why Sarah and the other women, even though they outnumbered the men, needed to watch their backs.

Sarah and the others limped westward through manzanita and digger pines, traveling over lower ridges and rounded hills, still aiming southwest as best they could reckon it. They were having a hard time walking normally now, staggering as if drunk at times and needing to stop to rest every quarter of a mile or so. When they came to logs or trees fallen across their path, they did not have the strength to climb over them. They simply put their arms on the obstructions, embraced them, rolled their bodies over the top, and dropped onto the other side. After about four miles of this, they came across the bloody footprints of Luis and Salvador. Then, two miles farther on, they came upon the two Miwoks themselves near a small stream.

John Sutter would later say that Luis and Salvador—his "good boys," as he called them—were gathering acorns when the snowshoe party found them, a not-unlikely thing for two starving young Miwok men to be doing under the circumstances. California Indians were the most omnivorous in North America. The Miwoks disdained few food sources, and Luis and Salvador would likely have seen gustatory and nutritional opportunities that Sarah and her companions would never have recognized as food. The Miwoks knew how to leach the tannic acid out of acorns by immersing them in sandy streambeds. They ate various kinds of grubs and larvae, earthworms, crawfish, snails, and freshwater mollusks. They ate a wide variety of native plants and roots. They knew how to grind up buckeye pods and throw them in a pool to poison fish. They knew how to construct simple snares to catch birds, rabbits, and other small mammals. Even spawned-out salmon carcasses or carrion were valuable to them—they ground up the vertebrae of dead fish and deer and made soft, mealy cakes that they could roast on the hot stones around a fire. They would eat the flesh of almost any mammal, except the domestic dog, which they considered the deadliest of poisons.*

Some accounts would later assert that Luis and Salvador were simply lying on the ground, dead, when the snowshoe party discovered them. But in fact they were alive, though greatly weakened. Unlike their white companions, they had not partaken of human flesh, and so except for whatever they might have gleaned from the countryside, they'd likely had little nourishment since December 21. The party moved a short distance past them and paused. Then William Foster took the flintlock gun and went back. He told Luis he was going to kill him and aimed the gun at the young man's head. Up the trail, Sarah and the others heard a shot. Then a long pause while Foster reloaded. Then another shot.

They stripped the flesh from the young men's bones and set them aside. Then they built a fire and dried the flesh while they roasted and consumed the more perishable parts of the bodies.

During the night it began to rain. By the next morning, Luis's and Salvador's severed heads and bloody

---

* If they followed the custom of the closely related Maidu, they also would not touch the meat of the grizzly bear, because the bear might have eaten human flesh and to eat of it would be akin to cannibalism, widely taboo among the Indians of Northern California.

bones lay in muddy pools of water. The snowshoe party packed what they could of the young men's dried flesh on their backs and pressed on. Their energy only modestly replenished by the lean meat they were now eating, they wandered through the rain, painfully and slowly. They saw deer regularly now, but Eddy was too weak to take steady aim at them. The country gradually opened out into more gently rolling ridges and knolls, and oaks began to replace the digger pines. It started to look something like what Sarah and Jay had dreamed California would look like. But in the rain, the mud, the grief, the fear, and the pain simply of walking, this California must have seemed a cruel, gray shadow of the one they had dreamed of. Sarah limped forward, concentrating for the moment on the simple task of putting one foot in front of the other.

Time began to collapse for them now. They wandered forward for another day or two. Then one of them found a human footprint, and then another and yet another in the mud. They gathered around the tracks and began to follow them. The prints led to what seemed to be an Indian trail through the brush. They hobbled along the trail as quickly as they could manage and then suddenly came around a chaparral-covered hillside and saw an Indian village laid out before them.

It might have been the Maidu village of Takema near modern-day Colfax. It may be that they had wandered as far south as the village of Hangwite near Auburn. Or it might have been another, unnamed Maidu village. Whatever its name, it consisted of a cluster of larger, round, earth-covered lodges called *k'um*, intermingled with smaller, conical brush shelters called *hübo*. Columns of smoke rose from chimney holes in the center of the larger lodges as well as from outdoor campfires all through the village.

The Maidu were astonished by the sudden appearance of seven pale and spectral figures on the edge of the village. They stood staring at them, mouths agape. The Maidu men wore simple, loose mantles of deer or rabbit skins around their shoulders, woven hairnets, and little else except simple moccasins. The women wore only aprons made of shredded bark.

As Sarah and her companions began to lurch toward the villagers with their arms outstretched, imploring aid, some of the Maidu at first ran away into the brush. But when they saw the condition that the white people were in, they approached them cautiously. Then they began to gather nosily around the whites, pointing and gesticulating at their hollow faces and gaunt frames. Some of the children cried and hid their eyes in horror.

Then they must have led Sarah and the others into the smoky interiors of the earthen shelters and laid them down on willow platforms covered with pine needles, tule mats, and soft animal skins. The Maidu, if they served the whites as they served themselves, used pairs of sticks to drop hot rocks from the fire into coiled willow baskets of acorn mush. When the mush was warm, they showed the whites how to eat it, scooping it out of the baskets with their index and middle fingers and then licking it off. Then they made patties of the mush, placed them between leaves, cooked them on the hot rocks surrounding the fire, and handed the hot acorn cakes to the strangers.

William Eddy could not manage to eat the stuff and settled for brewing tea from fresh grass instead. But Sarah and the rest of her companions ate the acorn mush and cakes, unpalatable though they probably found them. Then they lay on their backs and looked around and found themselves in circumstances that must have been stranger than they could ever have imagined, in huts full of odd smells, surrounded by dark faces peering back at them, speaking at them in an incomprehensible language, with the sound of rain splattering in the mud outside. But they were warm and dry for the first time in a very long time, and one by one they fell asleep.

So far as we know, none of them told the Maidu—close kindred to the Miwoks—about the meat still stored in their packs.

Over the next several days, it continued to snow in the mountains. Patrick Breen wrote in his diary, "Snowing fast wind N.W. snow higher than the shanty must be 13 feet deep don't know how to get wood this morning it is dreadful to look at." In the foothills it continued to rain, but the snowshoers, conscious that they still bore the burden of telling the outside world what was happening in the mountains, used signs to let the Maidu know that they wanted to continue to the nearest American settlement as soon as possible. They set off again with Maidu guides, following Indian trails through the brush. The going was slow, though. Even with their stomachs full, they were still malnourished and weak. But it was their swollen and bleeding feet that posed the largest problem now. They hobbled a hundred yards at a time, continually having to sit and rest in the relentless downpour.

They made four miles and arrived at another Maidu village, where they rested again, trying to regain strength and heal their feet. Then they resumed traveling, moving ever more slowly, following guides from one Maidu settlement to another. They were among broad-branched oaks now, almost on the edge of the Sacramento Valley; they could see it clearly from hilltops, stretched out before them, vast and flat and immensely fertile. The weather had cleared, but colder air had come in behind the rain. The mornings were frosty, the nights bone-chilling, and they were dressed only in tatters, some of them virtually nude.

By the morning of January 17, Sarah could barely walk at all. Despite their diet of acorn gruel, she and her companions suffered from a host of symptoms related to their months of malnutrition. As Sarah's body had tried to extract protein and vitamins from nonvital tissues, she had grown skeletally gaunt. She had almost certainly begun to experience bleeding gums from scurvy, yellowed skin from jaundice, intestinal bleeding, double vision, and slurred speech. As the nerve fibers in her extremities had begun to waste away, she likely had severe pain followed by loss of control in her arms and legs. Her skin was by now probably dry, scaly, and badly sunburned, her eyelids swollen with edema, her voice hoarse, her lips cracked, her tongue blistered. From time to time, she likely vomited green bile.

Finally, after stumbling on for two or three miles that morning, she simply gave out. She sat down by the trail and went no farther. Mary Ann, Harriet Pike, Sarah Foster, and William Foster did the same.

They had walked across the Sierra Nevada Mountains in the middle of winter, covering more than seventy miles of granite and ice and snow. They had been malnourished before they'd even begun. They had battled hypothermia every day of the trek. Much of the way, they had struggled through deep drifts of snow in heavy, clumsy, homemade snowshoes. Most of the rest of the way, they had been without intact shoes. They had walked as far as they could and come at last to rest on the very verge of what they had sought for so long. But now they could go no farther.

William Johnson's Ranch, the place toward which Sarah had been inevitably, though perhaps unknowingly, moving since she left home back in April of the preceding year, was not much of an establishment, less imposing even than the miserable stockade at Fort Bridger, the last American outpost any of them had seen, an eternity ago. The only structures were Johnson's house—a small two-room building, one half made of logs and the other half of adobe bricks—a few pens cobbled together from poles for the livestock, and a few rustic cabins that recently arrived emigrants had erected as temporary homes.

The land on which these structures stood was reasonably fertile, and Johnson and his partner, Sebastian

Keyser, farmed it, or at least oversaw the farming of it, most of the actual labor being supplied by several dozen naked Miwoks and Maidu. Johnson was fond of drink and generally disinclined to exert himself, but his property had the advantage of sitting squarely on the emigrant road leading down out of the Sierra Nevada and the even greater advantage of being the first American establishment that hungry and ill-provisioned emigrants encountered when descending that road.

In late October, just ahead of the snowstorms that closed the mountain pass to the Donner Party, several of the families that Sarah had first met across the Missouri River from St. Joe had arrived at Johnson's. Reason P. Tucker and his sons had arrived in October, after moving down from Bear Valley along with Matthew Ritchie's family and the Stark family. More of Sarah's traveling companions from back on the plains had also stopped at Johnson's but then moved quickly on to Sutter's Fort. A few had moved deeper even into California, but with winter closing in and the Sacramento Valley increasingly inundated by floodwaters, the Tuckers and the Ritchies had taken up winter quarters at Johnson's Ranch.

And it was at Johnson's Ranch that late on the afternoon of January 17, fifteen-year-old Harriet Ritchie saw two figures coming slowly down the flood-swollen

Bear River, approaching her family's cabin. She studied the men approaching her. One was an Indian, and
he was holding up the other, a white man—a gaunt
figure with rags hanging from his shoulders. The white
man's feet were bloody; his eyes were sunken into his
pale, bearded face. He was bent over as if by great
age. He stank of sweat and blood. When he got to the
door, he peered up into Harriet's face and whispered,
"Bread." She burst into tears.

Matthew Ritchie and his wife, Caroline, ushered
William Eddy into the cabin and laid him out on a bed.
Eddy croaked out that he was of the Donner Party and
that there were six more like himself, dying or dead,
some miles back up the trail. Caroline Ritchie brought
him food, and as she spooned it into his mouth, she,
too, began to sob.

Harriet ran from cabin to cabin informing the
other families. Nineteen-year-old William Dill Ritchie
helped his father to rustle up some supplies. Men and
women came running, converging on the Ritchie cabin
carrying bread, coffee, tea, sugar, blankets, whatever
they could think of. Five men—Reason Tucker, John
Howell, John Rhoads, Pierre Sicard, and the Maidu
guide who had brought Eddy in—hastily gathered together all the goods they could carry, strapped them
on their backs, and set out on foot among leafless oak
trees, following the turgid river back into the hills.

It was nearly midnight by the time they reached Sarah and the others, lying in the mud where they had collapsed that morning. They were all alive, although nearly naked and so malnourished and hypothermic that they were hardly able to sit up, let alone walk. But they ate ravenously the food that was offered to them; then they wept and cried out and begged for more. When it was given to them, they ate until they grew ill and vomited, the food overwhelming their shrunken digestive systems. Then they ate more.

All that night at Johnson's Ranch, women continued to bake bread and stew beef. On the morning of January 18, a second contingent of men—Matthew Ritchie, William Johnson, Joseph Verrot, and Sebastian Keyser—set out on horseback with the fresh supplies, trailing a string of extra horses behind them. Without an Indian to guide him, Ritchie followed Eddy's bloody footsteps for more than six miles through the brush. It took them all day to reach the survivors.

Late that night, moving through dark oak woods, Sarah and Mary Ann finally approached the Ritchies' makeshift cabin by the Bear River. What they remembered for the rest of their lives was not the cabin itself but rather the warm, yellow lamplight that shone out through loose chinking—light coming to them through the black night as if miraculously, beckoning them to come back in out of the cold, to the hearth of humanity.

The next day an Indian runner was sent splashing through the flooded lowlands west of Johnson's Ranch to the home of John Sinclair, the new, American-designated alcalde, or chief magistrate, of Northern California. When the runner returned, he carried on his back something that Sarah and her sister needed nearly as badly as food and rest—a large pack of clean, fresh women's clothes.

But it would take far more than clean clothes to heal Sarah's soul. In the days, weeks, and months that followed, the few white women at Johnson's Ranch, and later at Sutter's Fort, devoted themselves to nursing the survivors of the snowshoe party. One of those who tended to them was a young Englishwoman who had immigrated to California the year before, Eliza Gregson. When she first looked into the survivors' eyes, Gregson was startled by what she saw looking back at her, and she later marveled at it.

I shall never forget the looks of those people, for the most part of them was crazy & their eyes danced & sparkled in their heads like stars.

12

HOPE AND DESPAIR

The news that Sarah and her companions had carried out of the mountains—news of whole families
snowbound and starving—startled and horrified the
handful of American settlers living at Johnson's Ranch.
For many of those who had themselves so recently made
it through the mountains, a sense of shared purpose,
a sense of common humanity, and a sense of Christian
duty all demanded that they do something to help their
countrymen. But to launch any kind of successful rescue effort, they knew they would need far more resources than could be mustered at Johnson's squalid
establishment. They would need food, blankets, horses,
mules, and, above all, men willing to hazard their lives
in the high Sierra in midwinter. Those were commodities that they could hope to find only at Sutter's Fort,
some forty miles away. William Eddy scrawled out a

letter detailing the dismal situation in the mountains and imploring Sutter's aid, and Johnson sent it off to the fort by way of a foot courier.

The Tuckers and the Ritchies began to slaughter beef cattle and to dry the meat. They cut hides into strips with which to make snowshoes. William Johnson put some of his Indian laborers to work grinding wheat into flour in stone *matates.* The whites used coffee mills to grind still more wheat. George Tucker and William Ritchie and other young men rode through the surrounding countryside searching for enough horses and mules to pack the supplies into the mountains. And they waited for more men and supplies from Sutter's.

Meanwhile, Sarah and the other survivors lay abed in various cabins alongside the Bear River at Johnson's Ranch, stunned and desolate, trying to recover. Mary Ann—whose feet were so swollen and injured that she would not be able to wear shoes for three months— spent some of her time writing letters to her mother, hoping that the rescuers would carry them into the mountains for her.

By January 25, her twenty-second birthday, Sarah must have begun to fully absorb the fact that she was a widow. She had much to grieve for in the loss of the two principal men in her life—her husband and her father—and all that they had so recently been to her.

And she had new practical worries as well. When she had left her father's family and married Jay, she'd moved from one financial dependency to another. With both now dead, she faced the cold fact that she had no particular means of support. And at twenty-two she was no longer a child in any sense of the word, and no one would be likely to treat her as such.

The relationship between adults and children and the line demarcating the distinction between the two were shifting in the 1840s, as were many other aspects of life. In the world into which Sarah had been born in 1825, American ideas of who exactly was a child and who an adult, and how children should be treated, were still shaped largely by seventeenth- and eighteenth-century attitudes. In that world, children had for the most part been thought of as miniature adults. As a result, the treatment of children and the expectations placed on them had often been exceedingly harsh and unforgiving by modern standards. These attitudes had resulted in the widespread abuse of even very young children, particularly in Great Britain, where eight- and nine-year-olds were sometimes put in chains and harnessed to carts in coal mines and made to drag the heavy carts for as many as eighteen to twenty hours a day.

Many of these attitudes toward children came to America with the first English colonists. As early as 1619, hundreds of pauper children were abducted in England and shipped to Virginia, to be bound out to service on farms and in manufactories with no pay. And as the Industrial Revolution brought large-scale manufacturing to America's cities late in the eighteenth and early in the nineteenth centuries, the demand for cheap labor soon had urban children working long hours in abysmal conditions. In 1790, Samuel Slater, an early American manufacturing magnate, set the trend when he realized that he could run his factory most economically if he dispensed with hiring adults and employed only children, from seven to twelve years old, to staff his first factory in Rhode Island.

As Sarah and her siblings were growing up, town boys whose parents wished them to learn a trade were often sent away from home by the age of fourteen to serve seven-year apprenticeships. Country boys much younger than that worked alongside their fathers from dawn to dusk. Country girls like Sarah and her sisters also worked grueling hours on their farmsteads from the time they could chop kindling or tote a bucket of water into the house. And by thirteen or fourteen, many of them were married and running households of their own.

By the 1840s, though, forces were at work that would eventually begin to ameliorate the situation for children, both in Britain and in the United States. Popular literary works like Elizabeth Barrett Browning's poem "The Cry of the Children" and Charles Dickens's *Oliver Twist*—both widely read in America—began to change attitudes. Romantic notions about the natural innocence of children began to clash with the harsh reality portrayed by Dickens and meld with Victorian sentimentality. In 1842, Massachusetts limited the workday of children under twelve to ten hours. Six years later, in 1848, the state of Pennsylvania passed the first minimum-age law, outlawing the employment of children under twelve in textile mills.

In keeping with a burgeoning American emphasis on individualism and self-improvement, the emphasis in child rearing began to shift from breaking the child's will to developing the child's inherent capacity to make his or her own way in the world. To be sure, methods often remained harsh, and the widespread abuse of children in American factories and sweatshops continued largely unabated throughout much of the remainder of the century, fed by a steady stream of immigrant children from Europe. But at least change had been in the wind as Sarah was growing up.

And, of course, most emigrant parents in 1846 loved their children no less than do modern-day parents, and they were no less anxious about them when danger threatened, as their behavior in the Sierra Nevada was demonstrating that winter. They, as much as any of us who are parents today, wanted to turn their children out into the world well equipped to have happy and productive lives. But their frame of reference, their concept of what would best accomplish that aim, remained in the 1840s fundamentally different from ours. Few of us today can imagine our daughters in a wedding bed at thirteen, nor our sons sent away to work full-time at fourteen, but both were still comfortable ideas for most of Franklin and Elizabeth Graves's peers.

So when Sarah had left home a married woman at twenty-one, all the expectations of adulthood had already long since settled squarely on her shoulders. And when she arrived at Johnson's Ranch, all those expectations remained in force, despite the horror of what she had just been through. As she contemplated her new status, the one light that she could look to in the darkness of her inner world was the hope that her mother and younger siblings were still alive and that the cache of coins her father had hidden in the family wagon was still in her mother's safekeeping. Meanwhile, everything in her future depended on the men hurrying around outside making preparations for a rescue. So far, though, nobody had gone anywhere.

. . .

As Sarah lay in bed at Johnson's Ranch, her mother was fighting for her brothers' and sisters' survival in ways that illustrate just how fiercely many emigrant mothers were devoted to the welfare of their children. Elizabeth Graves and Margret Reed, struggling to keep their respective broods alive, were waging a low-level war. Margret Reed, living now in the Breens' cabin, possessed only bits of ox hide with which to feed her children, and these bits of hide were a half mile distant, stretched out on the roof of the double cabin where Elizabeth Graves was still living. Beyond scraps cut from them, she could rely only on whatever ox bones the Breens chose to offer her. From time to time, Peggy Breen slipped Virginia Reed small pieces of poor beef, but only rarely. And in the end Margret Reed knew that Peggy Breen, like any mother, would feed her own children before any others.

In Elizabeth Graves's half of the double cabin also, there was now nothing to eat but hides. On January 21 the Reeds' cook, Eliza Williams, waded through the snow from the Graveses' cabin, where she had been living, to the Breens' cabin. There she implored her mistress for a bit of beef. She could not digest the scraps of hide that Mrs. Graves provided her, she said. But with nothing but the same for herself and her children, Margret Reed could offer Eliza no help. Patrick Breen made a simple notation on the result of the encounter in his diary that same day: "Mrs. Reid sent her back to live or die on them."

On January 30 the struggle between Margret Reed and Elizabeth Graves reached a new pitch. Graves—a woman her former neighbors in Illinois remembered years later to be extraordinarily generous—seized the hides and various other goods that Margret Reed had left at her half of the double cabin. She dragged them into her side of the cabin and announced that she would not return them until Margret Reed paid her back for the cattle that Franklin had sold to her when they'd first become snowbound in November. Nobody recorded the words that followed between Margret Reed and Elizabeth Graves, but one can imagine. John and Edward Breen, Patrick's teenage sons, went to the Graveses' cabin later that same day to try to recover Margret's goods by force or by diplomacy, but when they returned, they had only two paltry pieces of hide to show for the effort.

At the Murphy cabin that night, a third mother, Levinah Murphy, with no resources to fight over, watched in despair as her seventeen-year-old son, John Landrum, ceased his delirious ranting, took a few last rattling breaths, and died. In the three full months since they had all become entrapped, he was the fourteenth member of the Donner Party to die, all of the dead, so far, male.

On February 4, under gloomy skies, fourteen men finally set out from Johnson's Ranch in an attempt to reach the emigrants at Truckee Lake. Reason P. Tucker and another of that year's newly arrived emigrants, Aquilla Glover, led the expedition—which in time would come to be called "the First Relief." Sixteen-year-old George Washington Tucker accompanied his father. Colonel Matthew D. Ritchie also went along, as did two newly arrived emigrant brothers, Daniel and John Rhoads, a young man named Riley Septimus Moutrey, several sailors, a German with the nickname of "Greasy Jim," a half-witted boy named Billy Coon, Jotham Curtis, whom James Reed had rescued from Bear Valley back in November, and a still-emaciated William Eddy.

They made their way up the emigrant road alongside the Bear River, traveling through a wet landscape of manzanita, sprawling oak trees, and spindly digger pines. The road was bad, and the horses and mules repeatedly got bogged down in mud, requiring the men to unload the animals each time and pull them out with ropes, then reload them. It rained intermittently at first. Then, two or three days out, the skies opened up and torrential rain slanted down in sheets. On February 9 they reached the snow line, and after four miles of leading the horses and mules through snow up to the animals' bellies, they made an encampment at Mule Springs at an elevation of 3,849 feet.

The next day Eddy, still too weak to once again assault the high country, turned back with the pack animals. He did not yet know it, but both his one-year-old daughter, Margaret, and his beloved wife, Eleanor, were already dead, their bodies lying in the snow outside the cabins at the lake camp. The rate of dying at the lake was accelerating rapidly now. Since February 1, in addition to Margaret and Eleanor Eddy, Amanda McCutchen's one-year-old daughter, Harriet, and Augustus Spitzer and Milt Elliott had all died.

Reason Tucker and Aquilla Glover cached a portion of the provisions at Mule Springs for their return trip and left the boys, George Tucker and Billy Coon, to watch them. As the remaining men started to climb higher, each of them carried roughly fifty pounds of supplies on his back, along with a blanket, a hatchet, and a tin cup.

On the morning of February 14, they faced the daunting prospect of making the steep climb out of

Bear Valley to Emigrant Gap and then beginning to work their way among the high granite peaks to the east. Three of the men refused to go farther. Tucker pleaded with them and offered five dollars per day out of his own pocket for those who would continue, but in the end only seven men started up the ridge.

They traveled in single file, each man taking a turn going out ahead of the others to beat a path through the snow, then falling back to the end of the line. Once they reached the top of the ridge, they followed a sinuous course, winding among trees and around the sides of peaks. Aware that they, like the snowshoe party, could easily become lost in this terrain, they set fire to dead pine trees that they came to along the way, both to serve as markers for their return trip and to show the route to additional rescuers who they hoped would soon be following them.

They went on like this for the next several days, until, on February 17, they camped just short of the summit overlooking Truckee Lake, where they again built a log platform on which to kindle a fire. They guessed the snow to be as deep as thirty feet here.

On February 18 the First Relief carefully descended the granite cliffs and crossed the frozen lake. Just before sunset they approached the woods where they had been told they would find the lake camp. The snow was, by their reckoning, about eighteen feet deep here, and they could see no sign of life. Daniel Rhoads described what happened next.

> We raised a loud halloo and then we saw a woman emerge from a hole in the snow. As we approached her several others made their appearance in like manner of coming out of the snow. They were gaunt with famine and I never can forget the horrible ghastly sight they presented. The first woman spoke in a hollow voice very much agitated and said, "are you men from California or do you come from heaven?"

The woman was Levinah Murphy, whose son John had died on January 30. In the past few days, she had watched first the baby Margaret Eddy, then Margaret's mother, Eleanor, and then Milt Elliott all die in her miserable cabin built up against the gray, cold boulder.

Reason Tucker and his men went quickly among the survivors, doling out small portions of food in the failing light of dusk. Tucker and his men were horrified by what they beheld. These people were walking cadavers. He and his men, he said, "cried to see them cry and rejoiced to see them rejoice." The picked-over bones of oxen and putrefied bits of ox hide lay about on the snow. Worse, human bodies also lay scattered about, half covered by snow or by quilts. To get the bodies out of the cabins, the women and children who now mostly inhabited them had had to dig inclined planes and drag them up and out. They had not then had enough strength remaining to bury them any deeper than under a light cover of snow.

Tucker pushed on to the Graveses' cabin a half mile to the northeast and arrived there in the dark at about 8:00 P.M. He found the double cabin, like the others, buried under more than a dozen feet of snow. But it was a clear night, and a crescent moon hung in the western sky. He could make out a column of smoke rising from a hole in the snow. When he called out, an emaciated Elizabeth Graves and her children crept up out of another hole. She scanned the faces of the men assembled in front of her in the moonlight and asked where Franklin was, whether he and Jay and Sarah had gotten through. Tucker did not have the heart to tell her the truth. He said that they had all arrived safely but that their feet were too frozen to allow them to return.* Elizabeth Graves didn't buy a word of it, though.

---

* Tucker had not brought Mary Ann Graves's letters to her mother for fear that any news of what had happened to the snowshoe party would demoralize those whom he hoped to lead out of the mountains.

If Franklin had not come back, it must be because something dreadful had happened. For now, though, she could only imagine just how dreadful.

The next day Tucker and some of his men continued on to the Alder Creek camp, where they found an even more dismal scene. Hunkered down in their tattered tents with only a single ox hide left between them, Tamzene and Elizabeth Donner had for days now been watching their twelve children, ages three to fourteen, slowly approach starvation. With Jacob Donner dead, George Donner bedridden with his infected arm, and most of the teamsters dead, the two mothers and Doris Wolfinger had had only three teenage boys—Jean Baptiste Trudeau, Noah James, and Solomon Hook— to help them with the heavy and relentless work of cutting wood and shoveling snow off the tents.

Tucker and Rhoads explained to the women that they had been able to pack in only enough supplies to last a few days, but that they hoped other rescuers would be arriving soon. In the meantime they needed to take anyone healthy enough to endure the hike out of here. And they needed to do it now, before anyone got any weaker.

Both mothers faced agonizing choices. Elizabeth Donner had four children who were simply too young to have a chance wading through the deep drifts of snow, and there were not enough adults to carry them all. Tamzene had three of her own in the same situation. What's more, Tamzene would not hear of leaving George Donner here to die without her. They picked out a few older children who they thought might be able to make it out—Elitha, Leanna, George Jr., and William—and handed them over to Tucker and his men. The rest of the children would have to stay with their mothers. The sixteen-year-old teamster Noah James would go, and so would Doris Wolfinger. But Solomon Hook would stay, and Jean Baptiste Trudeau was told he must, much against his will, stay to help cut wood for the women. The rescuers felled a large pine tree to give him a head start. They doled out small bits of beef and flour. Then, just hours after they had arrived, they led the children away from their desolate mothers and disappeared into the snowy woods.

The following day Elizabeth Graves had to make her own heartrending decisions. She still had four children aged eight or younger with her—eight-year-old Nancy, seven-year-old Jonathan, five-year-old Franklin Jr., and her infant, Elizabeth. She would have to stay in the mountains to care for them. But twelve-year-old Lovina and fourteen-year-old Eleanor would go. Elizabeth wanted Billy to stay, to cut wood and do the heavy work, but he was bent, he said, on getting over the mountains and bringing food back for his little brothers and sisters. Elizabeth relented and said he could go. If Franklin was dead, as she must have suspected, Billy was her only assurance that someone would at least try to return for her and the younger children. That night he chopped and stacked a fresh supply of firewood for his mother.

Peggy Breen and Levinah Murphy made similar decisions. For the most part, the older children would go and the younger would stay. But there were exceptions. Two-year-old Naomi Pike, whose widowed mother was waiting for her at Johnson's Ranch and whose infant sister had died just two days ago, would go, Tucker and Glover announced. The men would carry her all the way. Twenty-three-year-old Philippine Keseberg had already lost one child here and wasn't willing to watch another die. She would leave her hobbled husband and carry her three-year-old Ada as far as she could.

Margret Reed felt the same way. With no remaining resources of her own, reduced to eating the Breens' cast-off bones, she was going, and so were all her children, even three-year-old Thomas. As long as they could walk, they were going to get out of there, or die trying.

John Denton, the English gunsmith, like many of the single men, was weakening rapidly, but he, too, decided he would go.

. . .

And so on the morning of February 22, the men of the First Relief stood in the snow at the eastern end of the lake, surrounded by children and grief-stricken mothers with hollow faces who clutched their children and wailed and then finally let them go. They trudged off into the snow—twenty-two of them—a long procession mostly composed of children and large men.

They had not gone far, though, before three-year-old Thomas Reed and eight-year-old Patty Reed gave out. Tucker and Aquilla Glover told Margret Reed the two children would have to go back to the cabins. Months later Virginia Reed described the scene in her letter to her cousin Mary.

> *O Mary that was the hades thing yet to come on and leiv them thar. [We] did not now but what thay would starve to Death. Martha [Patty] said, well ma if you never see me again do the best you can. the men said thay could hadly stand it. it maid them all cry.*

When Patty and Thomas had been returned to the Breen cabin, the First Relief continued on its way, crossing the frozen lake, wading through deep snow.

Exhausted and famished by the effort, they eagerly ate the remainder of their dried beef that night, knowing that they had cached a good deal more up at the pass for the return trip. The next day they scaled the cliffs at the western end of the lake. Nearly all of the men carried children, but still there were many children who had to walk. Six-year-old James Reed Jr. was up to his waist in the snow, but somehow he kept going, clambering over ice-slick boulders and wading through deep drifts of powder. With every step, he told his sister Virginia resolutely, he was "getting nigher Pa and somthing to eat."

On the evening of February 23, they reached the pass and the platform of logs that they had built on the way in. But when they got there, they found trouble. When they had left their cache of food here, they had secured it from animals, probably by tying it in a bundle and suspending it from the end of a branch in a pine tree, out of reach of bears. But they had not reckoned on smaller and more agile animals like martens and fishers, who could climb to the ends of the pine branches. The cache had been ripped open, the food consumed or dragged away.

Alarmed, Tucker immediately announced that for the remainder of the trip everyone would be on short rations. From that moment forward, Tucker feared for all their lives. "Being on short allowances," he confided in his notes, "death stared us in the face."

On the morning of February 24, following the burned-out snags they had left behind on their ascent of the mountains, the First Relief resumed moving west. After a mile or two, though, John Denton began to fall behind. Reason Tucker held back, waiting for him, but the young Englishman had grown weak and snow-blind, and it was clear that he could not continue. The others could not wait for him. Tucker built a platform of pine saplings and kindled a fire on it for him. He sat the young man down, took an expensive coverlet from his own backpack, and wrapped him in it. Tucker assured him that help would arrive soon, though both men must have known that there was little prospect of that. Then Tucker walked on down the trail and rejoined the others. When Denton's body was eventually found, it had been half consumed by animals.

They traveled on down the Yuba River. Thus far they had been fortunate enough to have fair, warm weather, but the warm afternoon sun exacted a price, thawing the crust of the snow and making walking much harder. Three-year-old Ada Keseberg couldn't manage it at all. Her mother had carried her for as long as she could. Then John Rhoads had carried her, but by that evening Ada Keseberg was failing. In the bitter cold of that night, she died.

When it was time to push on in the morning, Philippine Keseberg, who had seen her only other child buried in the snow back at the lake camp, would not part with her daughter's body. She sat clutching Ada to her breast, would not relinquish her, would not move on without her. Finally, as most of the others began to trudge away, Tucker sat down with her: "I told her to give me the child and her to go on. After she was out of sight, Rhoads and myself buried the child in the snow best we could. Her sperit went to heaven her body to the wolves." Then they continued along the Yuba and camped that night under the granite knob now called Cisco Butte.

By the middle of the next day, February 26, their provisions had grown so short that for a midday meal they resorted to toasting rawhide shoestrings. But Glover and John Rhoads had gone ahead looking for signs of a second relief effort, and shortly after eating the shoestrings, Tucker and his band of survivors came across two men with packs trudging through the snow toward them. Glover and Rhoads had indeed found a second relief party a few miles ahead and sent the two men back with dried beef. Billy, Lovina, and Eleanor Graves and Tucker's other famished survivors sat down in the snow and feasted. And to make the meal all the sweeter for Margret Reed, the men who had brought the supply of dried beef had also brought news that filled her heart with joy. The Second Relief, now just a few miles down the trail, was headed by James Reed.

At about the same time, Reed learned from Glover and Rhoads that his wife and two children were alive just up the trail.

James Reed had lived an adventurous life since last he saw his wife. After failing to get through the mountains in his first effort to reach his family in October, he had returned to Sutter's Fort looking for help. But he could not have arrived at a more inopportune time—all the able-bodied American men in the area had by now rushed south to join John Frémont and fight against Mexican forces still resisting the occupation of California. Unable to procure sufficient men or supplies for another rescue attempt, Reed put aside any immediate plans to effect a rescue and went instead to the American-held Pueblo de San José, where he enlisted as an acting first lieutenant in a volunteer company. While waiting for military action and pondering his next move, he petitioned the new American alcalde of San Jose, John Burton, for a large tract of land in the heart of the Santa Clara Valley, the second land petition he had filed in California in a little over a month.

On January 2, near Mission Santa Clara, Reed and a small detachment of men under his command joined a larger force of about eighty-five, closed with roughly one hundred fighters under the command of Don Francisco Sánchez, and engaged in a brief skirmish. An hour later one or two Americans were wounded, five Californians were wounded, three Californians were dead, and the Californians had surrendered. Using his saddle as a desk, on January 12, Reed wrote exuberantly to Sutter about his role in the battle.

I have done my duty and no more; but I am still ready to take the field in her cause, knowing that she is always right. I tell you, my friend, many were the dodges I made with my head from the balls that whistled by me. . . .

After what came to be called, somewhat vaingloriously, the "Battle of Santa Clara," Reed rode north to San Francisco, where he petitioned first U.S. naval authorities and then the general citizenry of the burgeoning American hamlet for aid in rescuing the Donner Party. The citizens responded with more than a thousand dollars, the navy with a promise to supply matériel and logistical support. A navy midshipman, Selim E. Woodworth, agreed to lead the effort to transport goods into the mountains with a contingent of navy men.

Meanwhile, William McCutchen had made his way to the Napa Valley, where he, too, was gathering supplies for an effort to rescue his infant daughter, Harriet, and the others still in the mountains. On January 27 he wrote Reed, somewhat testily, "You had better come in haste as there is no time to delay." He wanted to get under way by February 1.

But it was February 19 by the time that Reed had rendezvoused with McCutchen, driven a string of horses to Johnson's Ranch, and begun slaughtering cattle and drying strips of beef. On February 21 some of the men who would make the attempt caught up with Reed and McCutchen at Johnson's. Among them was a familiar face—Hiram Miller, who had been a teamster for the Donners back on the plains before going ahead with the Russell Party. Selim Woodworth, they reported, had paused at Sutter's Fort, and they could not say when he would arrive.

The next morning Reed told Johnson to inform Woodworth that he had gone ahead without him and to follow on his heels as soon as possible. Then, with about two hundred pounds of dried beef, seven hundred pounds of flour, and more than two dozen horses,

Reed, McCutchen, and the men of the Second Relief left Johnson's Ranch and started up the Bear River.

Sarah's heart must have lightened at least a bit as she watched Reed, McCutchen, and their men leave Johnson's and ride up into the foothills of the Sierra Nevada. She knew now that there were two parties of strong, hearty, and well-provisioned men bent on saving what was left of her family, and she knew that a third, led by a U.S. naval officer, would soon be on its way. The First Relief, in fact, had been gone long enough that she could begin to expect—or at least to hope—to see them riding down the river with her mother and siblings any day now.

But she could not know just how dire the situation at the lake camp was now growing. On the same day that advance men from Reed's Second Relief met the returning First Relief and handed out food, some of the women who had been left behind in the mountains were making desperate and hitherto-inconceivable decisions. Patrick Breen recorded their plight in his diary the next day.

Mrs. Murphy said here yesterday that [she] thought she would commence on Milt & eat him. I don't think that she has done so yet, it is distressing. The

*Donners told the California folks that they [would]
commence to eat the dead people 4 days ago. . . .
I suppose they have done so ere this time.*

The following day, February 27, Reason Tucker and
the First Relief met James Reed, William McCutchen,
and the main body of the Second Relief somewhere in
the vicinity of Yuba Gap. Told the afternoon before
that his wife and two of his children were alive, Reed
had stayed up through the night, baking sweet cakes
for them.

When he saw black dots on the bright snow ahead
slowly growing larger, Reed rushed forward to meet
Margret, Virginia, and James Jr. Recognizing him,
they stumbled forward themselves, trying to close the
distance between them. Margret Reed fell to her knees
in the snow, overwhelmed with the emotion of seeing
her husband for the first time since he had ridden
away into the Nevada desert and an uncertain fate
five months before. Reed stooped to embrace her. But
when he looked into her eyes, and then into the eyes
of the other survivors, he was horrified, much as Eliza
Gregson had been when she first looked into the eyes of
the snowshoe party. "I cannot describe the death-like
look they all had 'Bread Bread Bread' was the begging
of every child and grown person," Reed later wrote.
"I gave to all what I dared."

Reed already knew from Glover and Rhoads that two of his children—Patty and Thomas—were still at the lake camp and that his journey wasn't going to end there. So he stayed with Margret only minutes before Tucker and Glover lead Reed's family and the other survivors down toward Bear Valley. Then he and McCutchen and their men began to hike eastward toward the lake camp, meeting stragglers from Tucker's group and handing out sweet cakes to them as they went.

Forty-eight hours later, Reed and the Second Relief crossed Truckee Lake and approached the Breen cabin. The snow had melted a bit since the First Relief had left, and the eaves of the cabin were now protruding above the snow. Reed saw his daughter, Patty, sitting atop the cabin, her feet resting on the snow. When she saw that it was her father who was approaching, Patty Reed ran toward him but fell face-first into the snow. Reed bent and scooped his daughter out of the snow and embraced her quickly. Then he plunged into the dank depths of the cabin, where he found his son, Thomas, alive but asleep on a bed of pine boughs. As Reed later reported to a writer named J. H. Merryman, he was that afternoon "in raptures."

The men of the Second Relief went next to the Murphy cabin. The scene there was ghastly. Skeletal, hollow-eyed women and children crawled out of the dark portal of the cabin and gathered around them.

Scattered about the cabin, where Tucker and Glover had seen bodies covered by quilts a week and a half before, human bones now lay, with shreds of pink flesh still clinging to them. Levinah Murphy took Reed down into the reek of the cabin where the surviving children in her care had been lying in bed for fourteen days, too weak to stir. Clumps of human hair were strewn about, matted in bloodstained clots.

Reed told the survivors that in just two days those who were able would have to walk out. One of the men set about making soup to build the strength of those who would make the attempt. Then McCutchen and Reed continued on to the Graveses' cabin.

They found Elizabeth Graves and all her remaining children alive but, like all the others who had been here since October, emaciated and nearly disabled by famine. Again they doled out small amounts of food and explained that in two days' time they would be leaving with all who were willing and able to accompany them. Then Reed went outside and joined William McCutchen to help him dig the body of his infant daughter, Harriet, out of the snow and rebury it in some exposed soil near the cabin.

The next morning, March 2, Reed, McCutchen, and several of the others pushed on to the Donners' camp at Alder Creek. A few of the men, though, stayed behind with Elizabeth Graves. She had told them of a problem that she needed help with. Somewhere beneath all the snow surrounding her cabin lay her family wagon and the cache of silver coins on which her family's whole future now depended. Under her direction the men began to dig for the wagon and the silver. But she could not have been happy that so many now knew about something she had thus far successfully kept concealed.

When they arrived at Alder Creek, Reed and McCutchen found two men they had sent ahead— Nicholas Clark and Charles Cady—tending to what was left of the Donner families. Elizabeth Donner, lying prostrate in her tent, was too enfeebled to do much of anything for her five children. In the other tent, George Donner lay near death. Tamzene was emaciated, but still healthy enough to get around. The three children remaining there—Frances, Georgia, and Eliza Donner—were all reasonably hale and hearty. Many years later Georgia would explain one possible reason that the children in the camp appeared more robust than the adults. For at least several days now, the children had been receiving more nourishment than that supplied by their usual rations of burned ox bones and boiled hides. Georgia sadly and cautiously recounted what the adults had decided to do.

When I spoke of human flesh being used at both tents I said it was prepared for the little ones at both tents. . . . I did not mean to include the larger children or the grown people because I am not positive that they tasted of it. Father was crying and did not look at us during the time, and we little ones felt that we could not help it. There was nothing else.

Reed and McCutchen quickly assessed the situation in the two tents and then told the two mothers there that once again they had to make some quick and difficult decisions, just as they had less than two weeks before. Anyone who could walk should come with them immediately, the men said. Those who could not should wait for Selim Woodworth and his men, who Reed and McCutchen believed would arrive soon. Elizabeth Donner was clearly among those who were too weak to walk out, so once again she had to decide whether and how to divide her children. This time she chose to send seven-year-old Mary, five-year-old Isaac, and her son by a previous marriage, fourteen-year-old Solomon Hook. She would not part with her two youngest, Samuel and Lewis.

The men urged Tamzene Donner to come with them and to bring all of her children. But Tamzene was adamant. Once again she would not leave her husband here to die alone. She and her three daughters would stay and wait for Woodworth. The men packed Mary and Isaac Donner onto their backs and set off for the lake camp.

They left behind seven days' worth of provisions and two men—Clark and Cady—to help Jean Baptiste Trudeau cook for the surviving Donners and perhaps nurse them back to health so that they would be able to travel when Woodworth arrived.

By eight o'clock that night, Reed and McCutchen and the others had made it, exhausted from carrying the two Donner children, back to the Graveses' cabin. McCutchen and Reed pressed on to the Breen cabin. There they finally sat down to a meal of fresh bread that Patty Reed had baked during their absence, using flour her father had left her that morning.

The night before, Patrick Breen had made the last entry he would make in the diary he'd begun back in November.

Fine & pleasant froze hard last night. There [were] ten men arrived this morning from Bear Valley with provisions. We are to start in two or three days & cash our goods here. There is amongst them some old [timers] they say the snow will be here until June.

And fine and pleasant as well, it must have seemed to Patrick Breen that last night, to be sitting in his cabin with the miracle of fresh-baked bread set before him, believing now that the horrors through which he and the others at the lake camp had been living were about to end. But for him, as for many of them, the horrors were in fact just about to begin.

HEROES AND SCOUNDRELS

On March 3, James Reed went to Louis Keseberg's lean-to shanty at the lake camp. The last time the two men had had anything to do with each other was back on the Humboldt River on October 5, when Keseberg had stood in the sand exhorting the others to get a rope and hang Reed from the yoke of his wagon.

Keseberg was now so feeble that he could do no more than to lurch about in the shanty. His face was shrunken, his thin brown beard long and scraggly, his clothes filthy and infested with lice. Reed took Keseberg's clothes from him. Then he got a bucket of water and warmed it at the fire and took a rag and bathed the man's reeking body. He combed his hair and dressed him in clean clothes. He gave him a bit of flour and about a half pound of jerked beef, all he could spare from his own pack. He told Keseberg that he would come back in two weeks and carry him over the mountain. Then he went out into the snow to join McCutchen and the other rescuers, who had been busy washing many of the children and clothing them in fresh flannel.

It was midday already, late to be starting out, but Reed wanted to make some progress toward the summit before nightfall. So the men of the Second Relief, after leaving one of their number, Charles Stone, to look after those who would remain at the lake camp, set off through the woods with seventeen survivors trailing behind them—this time three adults and fourteen children.

Many of them were better clothed than they had been in months—in addition to the fresh flannel clothes for the children, Reed had brought twenty-two new moccasins to replace the worn-out shoes of the emigrants. The adults all carried something, or someone. They carried some biscuits and a bit of jerked beef, perhaps a blanket or a quilt. Many of the men carried children whose parents already had their arms full with other children. Peggy Breen carried four pounds of coffee, a few strips of beef, a bit of tea, and a lump of sugar tied in a bundle at her waist so that her arms were free to carry her infant, Isabelle. Patrick Breen limped along carrying three-year-old Peter. No one was more burdened down, though, than Elizabeth Graves. She carried one-year-old Elizabeth and something that she knew that Jay would not want her to leave behind, the violin he had brought across the plains. And something else, even more valuable, and much heavier—the hoard of silver coins that some of the men had helped her retrieve from the floor of her family wagon.

Thus encumbered, slogging through slushy snow, they did not get far that first day. By late afternoon they had gone only partway up the length of the lake, and so they made a camp on a bare patch of ground on the lakeshore, just about two miles from the cabins. The weather had been warm and clear for more than two weeks now, and the members of the party counted their blessings as they contemplated the rate at which the snow was melting. That evening, as twilight faded to night, Patrick Breen took Jay Fosdick's violin and serenaded the others, the notes rising and falling plaintively as an almost-full moon rose over the cabins to the east. The spirits of many in the party began to lift for the first time in weeks.

The next morning, though, as the party prepared to push on, someone made a joke about Elizabeth Graves's coins, and whether the men should play a game of euchre to determine who should get them. Elizabeth was not amused. For her the bag of coins must have rapidly been becoming a cruel burden, both psychologically and physically. Essentially useless to her in her present circumstances, it was nevertheless the vessel in which all her hopes and the hopes of her children lay. Particularly with Franklin dead, as she must by now have divined he was, it represented the only form of financial security she had. But with the cliffs of the pass looming ahead and a small child to carry, the heavy coins also represented an encumbrance that might well mean the difference between living and dying. And surrounded by men—many of whom she did not know and some of whom were here primarily to make money—she had no real assurance that it would not be taken from her whether she lived or died.

As the rest of the party set off toward the western end of the lake, Elizabeth Graves hung back until the others were out of sight. She measured out a distance of about thirty feet from a large rock, scratched a shallow hole in a patch of bare earth, and buried the coins. Then, clutching her infant daughter, she hobbled ahead to join the others.

Once again they made only two miles that day and camped at the western end of the lake, under the forbidding granite cliffs that led up to the summit. Their evening meal was spare that night. Reed had grown alarmed at how little they had left in the way of provisions, so everyone was limited to a bit of gruel made from their remaining flour. On the morning of March 5, Reed calculated that he had only enough left for two scanty meals for each person, enough for breakfast and dinner that day, then nothing more until they reached their first cache. Late that afternoon they struggled across the summit and arrived at the remains of the camp that the First Relief had made on their way toward the lake. Tucker and Glover had left behind a platform of green logs on which to build a fire. The camp was located in an exposed spot at the eastern end of a long meadow just west of the pass.

During the day the skies had begun to grow overcast, then leaden. Now they were nearly black with storm clouds, and the temperature began to plummet. An iron cold began to lash the tops of the trees fringing the meadow. Reed and McCutchen set the men to cutting pine boughs for beds and building a windbreak, piling snow and more pine branches around the fire platform. They felled several trees in such a way that they toppled over and intersected near the platform, to provide a ready supply of firewood. With nightfall rapidly approaching, there was not time to do much more. Reed found time, though, to scribble notes for a journal entry.

Night closing fast, the Clouds still thicking terror
terror to many, my hartte dare not communicate
my mind to any, death to all if provisions do not
Come, in a day or two and a storm should fall on us.
Very cold, a great lamentation about the cold.

By sunset the wind began to howl through the peaks around them. Later that night, snow began to slice down out of the sky, plastering everyone white as they huddled around the fire.

More than a century and a half after the fact, historians and climatologists still debate whether the Donner Party fell victim to unusually cold weather in the Sierra Nevada during the winter of 1846–47. There is much anecdotal evidence to suggest that they did.

Back on October 30, John Sutter had noted snow in the western foothills and said that "it was low down and heavy for the first fall of the season." Aboard the U.S. naval sloop *Portsmouth*, anchored in San Francisco Bay for much of that winter, observers more than once noted snow on the hills surrounding the bay and, on one occasion, in San Francisco itself, both rare though not unheard of in the twentieth century. George Tucker, who spent the winter in the foothills of the Sierra, said that it had rained "nearly all winter and the country was all covered with water." Daniel Rhoads said, "This last winter is the coldest has ever been known in Calafornia." The next spring, eastbound travelers reported snow depths in the Sierra Nevada that today would be considered highly unusual so late in the season. Crossing the pass on May 1, 1847, just two months after Reed and the Second Relief became snowbound there, James Clyman reported drifts as deep as twenty or thirty feet near the summit. More than a month later, on June 7, John Craig encountered drifts still as deep as twenty feet.

There is anecdotal evidence, in fact, that the winter of 1846 was unusually cold across the Northern Hemisphere. At Fort Vancouver in the Oregon Country, the Columbia River was frozen over that winter. At their winter quarters in Nebraska, thousands of Mormons suffered terribly, and more than six hundred of them died, in bitterly cold blizzards that swept across the plains. Farther afield, on December 13, three days before Sarah and the snowshoe party departed the lake camp, Charlotte Brontë looked out the window of her father's parsonage in Yorkshire and wrote a friend,

The cold here is dreadful. I do not remember such a series of North-Pole days—England might really have taken a slide up into the Arctic zone—the sky

looks like ice—the earth is frozen—the wind is as keen as a two-edged knife.

The same bitter cold settled over Ireland that month, contributing greatly to the staggering misery and soaring mortality of desperate victims of the Potato Famine, a cataclysm that would in the following several few years claim perhaps a million lives.

In the Canadian Arctic, Sir John Franklin sat helplessly that winter on one of his two ships—the *Erebus* and the *Terror*—locked in ice at the south end of Peel Sound, off King William Island. Franklin, traveling south, had expected the passage to be ice-free when he entered it in September, as earlier explorers had reported, but he was quickly outflanked and entrapped by ice. What Franklin didn't know—but ice-core studies conducted in the twentieth century would show—was that he had sailed into Peel Sound just at the beginning of what would turn out to be a five-year-long period of exceptionally cold weather in the Arctic.

Franklin died the following June, his ships still trapped in the ice. His crew remained aboard the ships, dying one by one for another harrowing year until, in April of 1848, 105 survivors finally abandoned the ships. On King William Island, they converted one of the ships' boats to a fourteen-hundred-pound sled, piled supplies and personal possessions into it, and tried to escape overland. In the course of the next six miserable years every one of them died, wandering in the frozen wasteland, victims of lead poisoning from the canned food they were consuming, exposure, scurvy, and apparent cannibalism.

The scientific evidence for an exceptionally severe winter in the Sierra Nevada in 1846 is mixed. Tree-ring studies conducted in the 1980s by the University of Arizona suggest that it was a low-precipitation year—tree rings from samples taken at Donner Summit and downslope in the western Sierra Nevada do not show the kinds of growth that would be expected in a year of heavy runoff from a deep, wet snowpack. And yet stumps left behind by the Donner Party, presumably cut off a foot or two higher than the level of the snow, are known to have stood as tall as twenty-two feet, far above normal for the Donner Lake area.

As Mark McLaughlin points out in his book *The Donner Party: Weathering the Storm*, the explanation appears to be that light precipitation in the Sierra Nevada does not necessarily mean either warm weather or scant snow. In fact, it may well mean the opposite. Cold air creates light, deep snow—powder. An inch of precipitation may produce twelve inches of snow if the air is relatively warm. But the same inch of precipitation will produce as much as twenty inches of snow if the air is cold enough.

So it seems that while the Sierra Nevada did not have more storms than usual that winter, the ten major storms it did have were very cold and left very deep accumulations of snow. The first of them came early, at the end of October, and the last of them came late, in March, just as James Reed and the Second Relief arrived at the long meadow east of Donner Summit and set up camp.

And they could not, in many ways, have picked a worse place to camp. Surrounded on three sides by high granite crags—now called Mount Disney, Mount Judah, and Donner Peak—and located at the very crest of the Sierra Nevada, the landscape in which they were encamped is perfectly configured to trap massive amounts of snowfall. Open to the west, and thus to the full brunt of cold, Arctic storms blown in off the Pacific, the bowl-like landscape captures, in fact, an average of forty-one feet of snowfall per winter. That is why in 1938, less than a hundred years after the Second Relief camped here, Walt Disney chose it as the site for what is now the thirty-seven-hundred-acre Sugar Bowl Ski Resort. The snow here, at sixty-eight hundred feet, is dry, powdery, and copious even in a year of normal precipitation—more copious, in fact, than at any other ski resort in California. And that is why the Central Sierra Snow Lab, a high-tech, instrument-laden facility that studies the extreme meteorological conditions of the high Sierra, is located just down the road. For experiencing blizzards in the Sierra Nevada, this is the place to be.

All through the night of March 5, the storm that had caught the Second Relief near the summit continued to intensify. The party formed a circle around the fire, their feet pointing inward, lying close to one another. Elizabeth Graves held her baby, Elizabeth. Nancy, Jonathan, and Franklin Jr. lay close by. Peggy Breen clutched her own infant daughter, Margaret, to her breast, letting her suck, though Peggy's milk had ceased to flow some days before. Periodically she peeked under her cloak at the skeletal baby to see if she was still alive, surprised each time to find that she was. Next to her, Patrick Breen and four more of their children crushed up against one another.

William McCutchen and James Reed and some of the other men got up now and then to forage for firewood, but each time they did so, they had to go farther out into the icy, black void beyond the firelight.

As the night wore on, the radiant heat emanating from the fire, ablaze on the large platform of green logs, began to melt a wide hole in the snow under the logs. The platform and the fire slowly started to sink into the hole. Some of the men gave up on gathering firewood and instead began to pray. Reed, McCutchen, and a few others continued gathering wood and shoring up the berm, frantically trying to keep the wind and blowing snow from extinguishing the fire. McCutchen, returning from one of his wood-foraging trips, sat with his back to the fire trying to warm up, so numb that he was not aware his clothes had ignited until all four of the shirts he was wearing were burned from his back.

Reed, by now, had grown desperately concerned about the lives of his own two children as well as all the others in his charge. He later remembered that he watched helplessly as "the pitiless snow beat fiercely against their thinly clad and weak forms; their blood grew chill in their veins, and death, with glaring eyes, stared them in the face."

By morning the fire had melted a pit nearly ten feet across and perhaps ten feet below the surface of the snow. Counting heads, Reed and McCutchen discovered that five-year-old Isaac Donner was dead and already frozen stiff in his blanket.

The storm continued all through the day. Reed and McCutchen went to and fro, every ten or fifteen minutes, climbing out of the pit in search of wood, braving the wind that cut through their clothes like cold steel. They were almost entirely out of food now and their stomachs were cramped with pain. Peggy Breen began to weep, then to pray, and then to rage at the men—shaming them for being paid three dollars a day to save them and yet letting them all freeze to death like little Isaac Donner. Her lamentations grew even louder when her son John, sitting on a log sloping down into the hole, slipped and tumbled headfirst toward the center of the pit. McCutchen caught the boy and saved him from horrible burns, but a bit later seven-year-old Mary Donner slipped and badly burned one of her feet in the fire. Peggy Breen grew quieter and began to recite Catholic devotions.

They all dreaded the coming of another night, but it came nonetheless. Reed had begun to have trouble with his vision, and by nightfall he was so entirely snow-blind he could not even see the fire blazing before him. Now it was mostly up to McCutchen to keep the fire going. As the hours wore on, the snowfall began to taper off, but temperatures plummeted and the cold grew lethal. Reed later called the night "one of the most dismal nights I ever witnessed and I hope I never shall witness such. . . . Of all the praying and Crying I never heard nothing ever equaled it."

In the flickering light in the pit, Peggy Breen heard Nancy Graves call out to her mother repeatedly to come and cover her, but Elizabeth Graves responded weakly that she could not, that she was too tired. Then Elizabeth Graves's breathing grew irregular. She began to make sounds that alarmed Peggy Breen, unnatural sounds, she thought. One of the men got up and examined Elizabeth, shook the snow from her blanket and re-covered her. Elizabeth rolled over awkwardly to one side, her arm akimbo, and then did not move anymore. Peggy Breen waited a bit and then crawled over to her and found her already cold to the touch. Nancy Graves took her feeble baby sister into her arms and sat next to her mother's body. At eight, she was now the oldest member of her family still alive in the mountains.

By about noon the next day, March 7, the snow had stopped falling. Reed and McCutchen gathered their men together and talked about what they should do. Then Reed announced their decision: He and McCutchen and the other men would continue until they came to one of their caches or to Woodworth's party and then send someone back with food. He was taking his own children with him, he said, and he would also take fourteen-year-old Solomon Hook, who seemed to be up to the trek.

The Graves children and Mary Donner were clearly too feeble to go—most of them were too weak to even crawl out of the hole in the snow. But the Breens were not as malnourished as most of the others, and they seemed to be more robust. Reed and McCutchen tried to talk Patrick Breen into making the attempt, along with his family, but Breen would have none of it. He and his family would stay here and wait for relief, he insisted. Reed called his men to his side and made them witness Breen's decision. If Breen's family died, their blood was on Breen's own head and not Reed's, he said. The men cut three days' worth of firewood and then called for Solomon Hook to join them. Hiram Miller took Tommy Reed on his back, Reed took Patty by the hand, and they walked off to the west.

The trees that the men had felled when they'd first arrived had tipped into the hole and now projected upward out of it at awkward angles. In order to stay warm, fourteen-year-old John Breen climbed down one of the trees deeper into the pit. Then he cut steps for the others to help them descend. At the bottom of the pit, Nancy, Franklin, Jonathan, and the baby Elizabeth Graves huddled by the fire, along with the Breens and seven-year-old Mary Donner. At least they now had protection from the wind, but none of them had eaten anything in more than two days. Up on the rim of the pit, rigid and cold, lay the bodies of Elizabeth Graves and Isaac Donner.

The same day that Reed and McCutchen walked away from the miserable pit in the snow that would eventually come to be called "Starved Camp," Reason Tucker, Aquilla Glover, and the rest of the First Relief led their band of survivors along the Bear River, through rolling, oak-studded hills. At about three that afternoon, they finally arrived at Johnson's Ranch.

For the first time since she had left the lake camp on snowshoes more than two and a half months before, Sarah saw Billy, Eleanor, and Lovina. At some point that afternoon or evening, Sarah and Mary Ann must have faced the grim task of telling their younger siblings that their father was dead, along with Jay. But the younger children had better news to report. So far as they knew, their mother and the rest of their brothers and sisters were still alive in the mountains, and Selim Woodworth, James Reed, William McCutchen, and other good men were on their way to rescue them.

In the mountains two of the three hearty young men that Reed and McCutchen had left behind to care for the survivors had meanwhile reconsidered their willingness to stay. Not long after the Second Relief left,

Charles Stone had hiked from the lake camp to Alder Creek and talked to Charles Cady. They would be better off out of there as soon as possible, they'd decided.

When a horrified Tamzene Donner heard that Cady was about to abandon her and the rest of the Donner family, she struck a desperate deal with the two of them—for a good sum of gold, perhaps as much as five hundred dollars, they agreed to take her three youngest daughters over the mountains. Once again she stood outside her tent and wept as she watched men lead her children off through the snow, the last of them this time—six-year-old Frances, four-year-old Georgia, and three-year-old Eliza.

When Cady and Stone arrived at the lake camp, they deposited the three girls in the cold, dank, recesses of the Murphy cabin. The only other souls in the cabin were the feeble skeleton who was Levinah Murphy; her one-year-old grandson, George Foster; her emaciated eight-year-old son, Simon; and a gaunt, hollow-eyed, and increasingly desperate Louis Keseberg. Cady and Stone decided to take shelter elsewhere, likely in the abandoned Breen cabin.

By the time the blizzard that had pinned down Reed, McCutchen, and their party near the summit had finally blown over, Cady and Stone had again revised their plans. With so much fresh snow on the ground, it would be hard enough hiking even without small children to carry, they decided. They assembled their packs and headed out across the ice of Truckee Lake, leaving the Donner girls behind with Levinah Murphy in the cabin built up against a boulder. For the Donner girls, it must have been a harrowing moment—their hopes for escape suddenly dashed as they were left in a fetid cabin with a woman who appeared to be more dead than alive.

Two or three days later, Cady and Stone hiked through Summit Valley, just west of the pass, where they found a deep, wide hole in the snow with a blue curl of wood smoke arising from it. On the edge of the hole lay a dead woman and a dead child. The young men peered down into the hole. At the bottom, a ragged woman, a cluster of children, and one emaciated man lay on their backs, staring up at them with hollow eyes. Charles Stone and Charles Cady turned their backs on the hole and quickly resumed their way westward.

Cady and Stone overtook Reed and the Second Relief farther down the Yuba, and shortly after that they all came across Selim Woodworth and his men encamped in the snow near Yuba Gap. With Woodworth were William Eddy and William Foster. The two survivors of the snowshoe party had set out from Johnson's

Ranch several days before, no longer willing to wait for someone else to rescue what remained of their families. Eddy had already learned that Eleanor and his daughter, Margaret, were dead, but he still hoped to save his three-year-old son, James. Foster learned now from Reed that his son, George, was alive in the Murphy cabin.

Reed told Woodworth that he had left more than a dozen survivors in desperate straits ten or twelve miles to the east. He exhorted Woodworth to press ahead as quickly as possible to rescue them. Woodworth polled his men and asked if they would go forward. But Woodworth's men studied Reed's and McCutchen's stricken and haggard faces, and the corpselike survivors, and promptly announced that they would go no farther. Foster and Eddy pleaded with them and offered to pay any amount, but Woodworth's men calculated that Eddy and Foster were most likely destitute and unable to fulfill their promises. Reed spoke up, promising to make good on whatever Eddy and Foster offered. But the men were unmoved. Eddy and Foster, desperate, said they would continue alone, without provisions. Reed took them aside and told them it would be suicidal and finally got them to agree to retreat temporarily to Bear Valley until something could be worked out.

When they reached the valley, Woodworth again refused to go into the mountains himself, but he finally agreed to pay out government funds to the tune of three dollars per day to any man who would go, and to pay an additional fifty dollars to any man who would bring out a child not his own. Eddy separately agreed to pay fifty dollars to Hiram Miller, who had just come down with the Second Relief, to return with them. Foster paid the same amount to a second man, William Thompson.

Three more men also agreed to go on the terms offered by Woodward. One was a volunteer from San Francisco, Howard Oakley. Another was Charles Stone, who might have begun to feel guilty about leaving the Donner girls behind, or who might simply have wanted the additional money that hauling one of them down the mountain would fetch. And the third was one of the Graves family's traveling companions from back on the plains, the tall, bearlike John Schull Stark, Matthew Ritchie's son-in-law.

The following morning the seven of them set out for the high country. Even as they began the climb out of Bear Valley, though, Eddy's and Foster's hopes had already been dashed up in the mountains. Three-year-old James Eddy was dead in the Murphy cabin. And a night or two before, Louis Keseberg had taken one-year-old George Foster into his bed with him. In the morning the boy was dead. As Levinah Murphy and the three Donner girls looked on in abject horror, Keseberg took the boy's limp body from the bed, carried it to a wall, and hung it on a peg, like a piece of meat.

For nearly a week after Reed, McCutchen, and the rest of the Second Relief departed the makeshift camp at Summit Valley, the people they had left behind there struggled to survive in the pit that their fire had melted in the snow. The hole had grown deeper and wider until it was fifteen feet in diameter and twenty-four feet deep, reaching now all the way down to the bare earth.

As the days dragged on, Patrick Breen largely gave up on living. For the most part, he simply lay listless on the muddy ground, staring up at the circle of sky above him. But Peggy Breen struggled to nurture the nine children there—her own five, Mary Donner, and three of Elizabeth Graves's orphans. The number of Graves children needing care had diminished by one shortly after Reed and McCutchen had left, when five-year-old Franklin Ward Graves Jr. died. The Breens had dragged the boy's body up out of the pit and laid it in the snow near his mother's and Isaac Donner's bodies.

For the first few days, Peggy Breen brewed small amounts of tea from her diminishing supply of tea leaves and doled it out to the children, to warm them more than nourish them. She rationed out small bits of sugar from the lump she had carried up the mountain, and dispensed some seeds she had also brought along. Every few hours, day and night, she or Patrick or their oldest son, John, crawled up one of the trees that had fallen into the pit and stumbled through the nearby woods searching for downed firewood. Each time they scanned the horizon for signs of rescuers, but each time there was nobody to be seen, no motion save the stirring of pine trees in the wind. Patrick Breen slid further into despair.

The weather was fair and relatively warm, and during the days some of them crawled up to the edge of the pit and sunned themselves, storing up warmth for the nights ahead, averting their eyes from the bodies that lay in the snow staring vacantly at them. But the nights were brutal. Without cloud cover, the warmth of the day was radiated quickly back out into the black void of space. As they lay in the pit, staring at white sheets of stars spread across the opening above them, they shivered and shook convulsively, aching with the pain of the cold. In the mornings their thin blankets and clothes were board-stiff, crusted over with a thick white rime.

As the week wore on, the sugar and the seeds and the tea began to run out, and finally there was nothing at all left to alleviate the stabbing hunger cramps of the children in the pit. All of them were profoundly emaciated, and with no body fat to insulate them, their internal temperatures hovered near the hypothermic range. Finally seven-year-old Mary Donner, the toes of her feet blackened by frostbite and the burns she had suffered after falling into the fire, could not stand the hunger pangs any longer. She suggested that they eat the dead.

Several days later, at about four o'clock in the afternoon of March 12, William Eddy, William Foster, and the rest of what was now the Third Relief trudged up the length of Summit Valley on snowshoes. At the far end of the valley, they could see a large, dark void in the snow from which was emanating a column of wood smoke. As they approached the column of smoke, they saw that there were bloody bones strewn around the lip of the crater. When they got closer, they saw what appeared to be a woman's body lying in the snow. It was hard to tell, though. Elizabeth Graves's body had been stripped of much of its flesh. The heart and the liver had been cut out of her chest and abdomen, and her breasts had been cut off. The rest of the bones were small ones, children's bones.

Down in the pit, a circle of living children, pale and skeletal, sat around a fire. For several days Patrick Breen had been bringing them bits of meat to roast on the fire. One of the children, eight-year-old Nancy Graves, did not yet know that the flesh she had been eating was her mother's—a revelation that when it came would so devastate her that it would lead to bouts of sudden, intermittent sobbing in her childhood and a sense of guilt from which she would never entirely recover.

Nancy Graves's later emotional distress was just one small thread in a much broader fabric of mental anguish that inevitably afflicted many of the Donner Party survivors and rescuers alike for years following the disaster. Even as Sarah and the other rescued emigrants at Johnson's Ranch waited to learn whether their loved ones would ever emerge from the mountains alive, silent and sinister processes were at work within them, processes that in many cases would transform the way they viewed and experienced the world for the rest of their lives.

Not all disaster survivors suffer from posttraumatic stress disorder (PTSD) and related syndromes, but large numbers do. Overall, 25 percent of people subjected to a traumatic event develop PTSD, but that number can more than double to 59 percent or higher among survivors of disasters. The precise rate varies according to a variety of factors, many of which were stacked against Sarah and her sisters. For one thing, females are four times more likely than males to develop the disorder when subjected to the same trauma. Children and young people suffer at higher rates than do more mature adults. Indeed, 100 percent of children who witness the homicide of a parent develop PTSD. And Sarah and her siblings had also experienced a witches' brew of additional traumas, each of which raised their risk: anticipation of suffering or danger to come, close exposure to dead bodies, witnessing the death of a loved one, the long duration of an ongoing trauma, a clear threat to their own lives, the loss of home and hearth, and experiencing physical pain or injury.

None of this, of course, even begins to touch the particular trauma that Sarah, Mary Ann, and Nancy Graves had suffered, along with a number of other Donner Party survivors—that of having eaten and seen eaten the flesh of their companions. Regardless of the necessity of having done so, they had violated a fundamental human taboo, and it was almost inevitable that they would experience significant amounts of guilt and its close cousin, shame. The two are not quite the same thing: Guilt revolves around feeling bad for what you have done; shame is feeling bad about yourself as a person because of what you have done. Guilt can actually be therapeutic, because inherent in the emotion is the idea that you can change your behavior and end the problem. Shame is a far more toxic emotion, because it implies that your character has been polluted by your actions. Deep-seated shame typically leads to a variety of anger-related emotional problems, particularly hostility and aggression.

Psychiatric researchers have only recently begun to understand that traumatic stress produces not just psychological changes but physical changes in the body, particularly in the brain. The hippocampus—the brain structure responsible for regulating memories and putting information into context—shrinks by as much as 8 percent, some of its cells killed by an excess of stress hormones such as cortisol. Under stress the amygdala—an inch-long, almond-shaped structure responsible for regulating emotions—becomes overactive, lighting up with activity like a pinball machine. The levels of neurotransmitters that regulate nerve impulses in the brain become unbalanced: Serotonin levels plunge; norepinephrine levels soar.

All this results in a kaleidoscope of psychiatric symptoms that plague victims of PTSD: panic attacks, flashbacks, dissociation (in which the patient feels detached from his or her own body), phobias, irrational avoidance of anything imagined to be related to the trauma, sexual dysfunction, eating disorders, emotional numbness, intrusive thoughts, and in extreme cases even psychosis and suicide. Among children there is sometimes a tendency to see "ghosts," apparitions that are apparently hallucinations brought on by the disorder. And to round out the misery, the vast majority of PTSD victims simultaneously suffer from at least one other psychological disorder, a full 80 percent experiencing depression and/or substance abuse.

The toll that traumatic stress takes is not confined to the brain and the psyche, though. Along with a host of other problems, chronic, recurring stress of the type associated with PTSD suppresses the immune system, rendering the entire body vulnerable to a variety of infectious agents. Researchers have for some time also noted high incidences of mortality from coronary heart disease among disaster survivors. It now appears that this is due, at least in part, to elevated cholesterol and triglyceride levels brought on by hormonal changes at work in the bodies of disaster survivors.

To what extent Sarah and her siblings suffered from any stress-induced symptoms we can never know with certainty. There was no formal knowledge of these syndromes in their time. No one had recourse to the wide variety of treatments available today, ranging from recently designed psychotropic medications to hypnosis, to eye-movement desensitization and reprocessing, to elaborate cognitive-behavioral therapies and carefully controlled reexposure techniques. For the most part, people were expected to keep their emotional problems to themselves. If Sarah suffered, she, like most of her fellow survivors, did so largely in silence and mostly unaware of why they were experiencing the problems they were. But Mary Ann Graves gave some insight into what she and likely her sisters were all experiencing many years later, when she said sadly, "I wish I could cry but I cannot. If I could forget the tragedy, perhaps I would know how to cry again," encapsulating even in those few words two of the principal symptoms of PTSD—recurring recollections of the trauma and emotional numbness.

At the crest of the Sierra Nevada, the men of the Third Relief were stunned by what they had found at the pit in the snow and uncertain what to do next. It seemed unlikely that any of these people, except perhaps Peggy Breen and her eldest son, John, could walk out of the mountains unaided, but Peggy Breen announced that she would not leave without her husband

and all of her children. Eddy and Foster wanted to push ahead immediately to the lake camp to look for their young sons. Charles Stone and Howard Oakley, there only as hired hands, wanted to return to Bear Valley as quickly as possible with the three surviving Graves children and Mary Donner. That, though, would mean leaving all of the Breens behind to wait for yet another relief effort, or for those who were going on to the lake camp to return.

The men stood in the snow discussing it. Finally a vote was taken. All except for twenty-year-old John Stark were for leaving the Breens. When his name was called, Stark stepped forward and said, "No, gentlemen. I will not abandon these people. I am here on a mission of mercy, and I will not half do the work. You can all go if you want to, but I shall stay by these people while they and I live."

Early the next morning, the party divided. Before dawn, Eddy, Foster, Miller, and Thompson resumed traveling east toward the pass and the lake camp. Charles Stone picked up the emaciated baby, Elizabeth Graves. Howard Oakley picked up the shrunken frame of Mary Donner, whose feet were too badly burned to allow her to walk. John Stark took charge of all the rest, placing Jonathan Graves on his back among his blankets and other gear and then leading Nancy Graves and the Breens step by step down the length of Summit Valley, traveling westward. As the smaller children grew exhausted from floundering through the snow, they took turns climbing onto John Stark's broad back, sharing the ride with Jonathan.

Eddy, Foster, Miller, and Thompson arrived at the lake camp before noon. Eddy and Foster hurried to the Murphy cabin, where they had last seen their sons alive on December 16. The two men rushed into the dark cabin and found a group of spectral figures crouched in the corners and lying on beds of pine boughs. In the dim light, they could make out the three youngest Donner girls—Frances, Georgia, and Eliza. Levinah Murphy was there, too, gaunt, wild-eyed, and disheveled. Her son Simon was there. And a startled and feral-looking Louis Keseberg was there as well. But the two boys for whom Eddy and Foster had come searching were nowhere to be seen. Eddy confronted Keseberg and demanded to know what had become of them. Keseberg told him flat out—the boys had died and been eaten. Enraged, Eddy threatened to kill Keseberg then and there, but the man was so emaciated and frail-looking that he resolved instead to wait until they got to California to commit the deed.*

---

* Eddy later attempted to make good on his threat in San Francisco but was stopped by James Reed.

Eddy and Foster were in no mood to linger in the presence of Keseberg. But as they tried to figure out what to do next, Simon Murphy noticed a haggard woman wandering through the woods as if in a daze. It was Tamzene Donner, come in search of her daughters. When she was brought into the cabin, her girls threw themselves into her arms and kissed her, but she was distracted, confused, not sure what was happening here.

The men, worried about another storm cutting them off on the pass, went outside to discuss how to proceed. When they reentered the cabin, they announced abruptly that everyone who was able to travel had to leave now. Almost as soon as they had met, Tamzene and her daughters were torn apart again. Hiram Miller picked up Eliza, Eddy picked up Georgia, Thompson picked up Frances. Foster picked up Simon Murphy. Levinah Murphy could not bear to see her youngest son go. Lying in her bed, she rolled over to face the wall. The men carried the children up out of the cabin to the open snow and hurriedly bundled them in warm clothes.

Tamzene Donner was frantic. Eddy and Foster urged her to come with them. She begged them to give her time to return to Alder Creek to see if her husband was still alive, but they would not chance lingering here another night. Jean Baptiste Trudeau and

Nicholas Clark, on whom she had largely depended, were determined to leave with the Third Relief. As the men carried her children away, Tamzene cried out to them, "Oh, save, save my children!"

They trudged off through the woods and onto the ice of Truckee Lake. At Alder Creek, Elizabeth Donner was dead or soon would be. Only four living members of the Donner Party would be left behind in the mountains—George and Tamzene Donner, Louis Keseberg, and Levinah Murphy.

Eddy and Foster caught up with and passed the remainder of the Third Relief several days later. John Stark, still carrying one or more of the smaller children on his back at a time, was moving slowly, leading Nancy Graves and the Breen family as they hobbled down the Yuba River. Later the same day, they passed Stone and Oakley, carrying the baby Elizabeth Graves and Mary Donner. The next morning they found Selim Woodworth and his men still encamped on the Yuba. When they told Woodworth that there were still four people in the mountains who ought to be rescued, Woodworth again declined to attempt an immediate rescue. His priority, he said, was to return to Mule Springs and arrange transportation to Johnson's Ranch for those who had already been brought out.

Two or three days later, Stark, the Graves children, and the rest of the Third Relief and their evacuees staggered into the camp at Mule Springs, where they found a pack train full of supplies, replete with all the food they could want.

After a few more days of riding mules and horses over muddy trails, they finally arrived at Johnson's Ranch late at night. It was the next morning before the new arrivals could see what they had striven to reach ever since leaving Illinois the previous spring. Many years later John Breen, who had just turned fifteen when he arrived, recalled that first California dawn.

It was early morning, the weather was fine, the ground was covered with fine green grass, and there was a very fat beef hanging from the limb of an oak tree, the birds were singing from the tops of trees above our camp and the journey was over. I [kept] looking on the scene and could scarcely believe that I was alive. The scene from that morning seems to be photographed on my mind.

But there were hard things that had to be said, and hard things that had to be accepted. William Eddy had lost his entire family. William Foster had to tell Sarah that their only child was dead. Eight-year-old Simon

Murphy had to tell his ten-year-old brother, William, they would almost certainly lose their mother. Nancy Graves, just eight, had to tell Sarah and the rest of her older siblings that their mother was dead, that their brother Franklin was dead, and that all their money was lost. There was so much anguish in William Johnson's two-room adobe that day that little Eliza Donner, who had just arrived with the Third Relief, had to flee the house to escape the heartrending scenes.

And no one could have been more devastated than Sarah by the time the sun rose that day. Everything for which she and Jay had wished, and almost everyone she had ever depended on, were now irretrievably gone. Almost everything she'd had to fear as she lay recovering at Johnson's was now in fact unfolding—she was suddenly both penniless and the titular head of a family of seven younger siblings, the youngest of whom was an infant girl who looked for all the world like a toy skeleton.

PART FOUR

In the Reproof of Chance

In the reproof of chance
Lies the true proof of men.

—WILLIAM SHAKESPEARE,
Troilus and Cressida

SHATTERED SOULS

Over the next six weeks, the last few horrific scenes of the Donner Party tragedy played themselves out high in the Sierra Nevada.

On April 13 one final expedition left Johnson's Ranch for the mountains. This one was not so much a relief party as a salvage operation, though it would come to be known as the Fourth Relief. The two Donner families were believed to have been carrying a large amount of gold and silver with them, as well as manufactured fabrics—calicoes and linens and silks that were precious commodities in California—jewelry, books, and other valuable goods.

William "Le Gros" Fallon, a mountain man of prodigious size, led the expedition. He and two others—Joseph Sels and John Rhoads—were to undertake the operation under terms and conditions set out by John Sinclair, who was acting both as the local alcalde and also as protector of the Donner children's interests. Since both Elizabeth and Jacob Donner were known for a fact to be dead, Sinclair decreed that Fallon and his men were to receive half of whatever booty was found, the other half to go to their orphans. If George and Tamzene Donner were found alive, the party was to negotiate directly with them for the terms on which they and their property would be extracted from the mountains. If they were found dead, the three men were once again to receive half the proceeds, to be apportioned by Sinclair, with the other half going to the Donners' orphaned children. Once again William Foster went, presumably to salvage what he could of his own possessions before someone else did. And Reason Tucker also decided to go along, as much for humanitarian as for financial reasons.

On April 17 the Fourth Relief reached the cabins at Truckee Lake at a little after noon. No one there was alive. Reason Tucker, who had already seen bodies lying in the snow on his previous visit, was nevertheless shocked by the charnel houses that the cabins had become since he had last been there.

Death & Destruction. Horrible sight. Human bones. Women's skulls sawed to get the brains.

Better dwell in the midst of alarm than to [remain]
in this horrible place.

Eleanor Eddy's and Levinah Murphy's partially butchered bodies lay in the snow, as did the remains of others
whom the men could not identify.

With nothing they could do for the dead, and no one
here living, Fallon and his men rummaged through the
cabins for the next two hours, working in the stench of
death, searching for valuables but finding little worth
packing over the mountains. Then they set off for the
camp at Alder Creek.

Along the way they came across the fresh tracks of
a lone individual who had recently traveled through
the snow, apparently moving away from Alder Creek.
When they arrived at Jacob Donner's tent, they again
found no one alive. But they did find trunks that
had been broken open and goods that had been scattered about in all directions—bales of fabric, shoes,
and schoolbooks. And once again they found bits and
pieces of human beings. In a kettle inside the tent, they
found what they took to be chunks of human flesh
cut up into serving sizes. Nearby they found George
Donner's severed head, his skull split open and the
brains removed.

They ransacked both Donner tents looking for the
stash of gold and silver coins that George and Tamzene

Donner were thought to have brought with them. Failing to find it, they began packing up the most valuable of the other goods. Then they camped for the night.

In the morning Foster, Rhoads, and Sels returned to the lake camp, trying to follow the mysterious tracks in the snow they had come across the day before. When they arrived at the cabins, they were astonished to find Louis Keseberg, alive, lying among a heap of human remains next to a pan of brains and liver.

The men demanded to know where Tamzene Donner was. She had been in good health when the Third Relief left the lake camp on March 13. She was dead, Keseberg replied. Dead and eaten up. They asked where the Donners' money was. Keseberg said he knew nothing of any money. But the men tore open a bundle in the cabin and found silks and jewelry and a brace of George Donner's pistols. They searched Keseberg's person and found $225 in gold coins in his waistband. They took the money, and then they began to threaten Keseberg, telling him he would hang for this in California, pressing him into a corner and demanding again to know everything—what had happened to Tamzene, how he had come to have so much gold, what explained the tracks in the snow, where the rest of the Donners' money was.

Keseberg poured out his version of events—that Tamzene had come to his door one cold night after George Donner died and said that she was going to go over the mountains alone, to see her children. She had fallen into a creek on the way to the cabins and become chilled and died that night, Keseberg said. And he said that before she died, she'd told him where her money was hidden at Alder Creek and that he'd promised to get it and carry it to her children in California. It was he who had made the tracks, after spending a night in the Donners' tent and retrieving the gold. He knew nothing about any more money than that, he said.

Suspicious, disgusted, and frustrated, the men left Keseberg alone and returned to Alder Creek, where Tucker, Fallon, and a sailor named Ned Coffeemeyer had remained to pack up the Donners' goods. The following morning all of them returned to the lake camp, carrying as much as they could of what they had salvaged there. By now they were convinced that Keseberg had murdered Tamzene Donner for food and concealed the rest of her money.

This time they were rougher with him. Fallon told someone to get a rope. He formed a noose and looped it around Keseberg's neck and threw him to the floor. Then he began to tighten the noose. Keseberg gasped and choked and finally cried out that he would show them where the money was if only they didn't murder him. Fallon loosened the rope, and after much delay Keseberg led Tucker and Rhoads off into the snow toward Alder Creek. The next morning they returned with $273 in silver that Keseberg had buried near the tents.

That afternoon the party set out for the return trip to California. They had more goods than they could carry, so they moved ahead by relays, each man carrying a bundle forward, depositing it on the ground, then returning for another. Keseberg, still lame from his foot injury back on the plains, limped along behind them as best he could and arrived late in camp each night. Several days out, as he prepared to make a cup of coffee, Keseberg noticed a bit of calico sticking out of the snow. Curious, he dug deeper and grabbed hold of something cold and solid. He tugged hard, and out of the icy tomb in which Reason Tucker had laid her two months before he pulled the frozen, blue-faced body of his daughter, Ada.

A few days later, the Fourth Relief rode into Johnson's Ranch on mules laden with salvaged goods, and the long, cruel saga of the Donner Party was finally over. But it had taken a terrible toll. In the end, of the eighty-seven people who emerged from the Wasatch

Mountains as official members of George Donner's company, forty-seven had died as a result of the tragedy. The toll had fallen disproportionately on the males in the company.

Of course, women in general now outlive men in North America and most of the rest of the world—by an average of about 5.3 years for girls born in 2003 in the United States.* It has not always been so, though. The greater longevity of women that we now take for granted in the United States is actually a trend that has emerged only in the last century and a half. After remaining largely unchanged for centuries, the human life span began to rise sharply in the 1840s, just as Sarah and her companions made their way across the country. While both genders benefited, the rate of acceleration was particularly dramatic for females, and by the end of the nineteenth century women began to open a gap that they have never surrendered. The average life expectancy for American women like Sarah, born in 1825, was well under forty; it has now more than doubled. Men now die earlier largely as a result of high

* Just how astonishingly lucky we are to live in the twenty-first century is underscored by a quick look at historical longevity rates worldwide. The average citizen of the Roman Empire could expect to live to about twenty-five. By the beginning of the twentieth century, the average global life span was still only about thirty-six years. But by 1995 it had reached sixty-five, and by 2008 in the United States it was seventy-eight.

rates of smoking, homicide, suicide, ischemic heart disease, war, and higher degrees of risk taking. Interestingly, men who have been castrated have been found to have a life expectancy 13.6 years longer than that of intact males.

In the world Sarah was born into, though, North American and European women could still expect to live only about as long as men. This was true, however, just among ordinary populations living under ordinary circumstances. The numbers were different then, as they are now, when those looked at are individuals living under extreme conditions, such as famine and life-threatening cold. In particular situations where men are able to take more than their fair share of resources, and inclined to do so, they sometimes do outsurvive women. But when scarce resources are shared more or less equally, the opposite is very much true.

Donald Grayson at the University of Washington has studied this phenomenon in relation to the Donner Party and come up with some interesting, and for those of us who are male, discomfiting, observations. Overall, Donner Party men died at nearly twice the rate of women (56.6 percent of the males, 29.4 percent of the females). They died much sooner, too. Fourteen Donner Party males died before the first female did. And it was men in their prime years who died earliest

and in the largest numbers. Of twenty-one men between the ages of twenty and thirty-nine, 66 percent
died; of thirty women in the same age group, only 14
percent died.

Grayson's study implicated interesting factors in the
differing mortality rates for males and females. Women,
of course, generally have smaller bodies, and one would
therefore expect that they would lose their core body
heat to the outside environment more quickly than men.
But in fact they actually retain core body temperatures
better. This is partly because women have higher proportions of body fat and a higher proportion of that fat
is located subcutaneously (just under the skin). This,
in effect, provides them with a layer of insulation and
a survival advantage in extreme cold. Women also, on
average, maintain lower skin temperatures than men,
resulting in less temperature differential between the
skin and the environment and therefore less heat loss.

Grayson also noted that men's larger muscle masses
burn larger numbers of calories than women's do,
simply in the ordinary business of moving about, let
alone in doing the kinds of things that the men of the
Donner Party had been forced to do. Long before they
reached the Sierra Nevada, particularly in cutting a
road across the Wasatch, the men had burned up much
of the energy reserves stored in their bodies. Then they

had burned up most of whatever was left by doing the largest share of the heavy work of building and maintaining the camps at the lake and at Alder Creek.

Perhaps the most interesting of Grayson's findings, though, was that, male or female, those who traveled with a large family group had a better chance of survival than those who were on their own. This is in keeping with other studies correlating survival with the size of social networks. Scientists are not sure why this effect takes place. Theories point to better sharing of critical information and scarce resources, better mutual aid in emergencies, better emotional support, and the possibility that the immune system is physically stimulated by close proximity to loved ones. At any rate, mortality in the Donner Party followed the trend, and again it worked heavily against the males. There were fifteen unattached men between the ages of twenty and forty in the Donner Party. Only three of them survived.

As the survivors began to recover, many of them faced the task of letting the folks back home know, at least in part, what had happened. On May 16, Virginia Reed sat down and wrote what was probably the most heartfelt letter home from California that spring, addressed to her cousin Mary Keyes. As she moved toward the close of the letter, after detailing her family's sufferings, Virginia wrote in an unschooled way words that have since become emblematic of the entire Donner Party story.

> *O Mary I have not wrote you half the truble [we have had] but I have Wrote you anuf to let you [k] now what truble is. . . . Don't let this letter dishaten anybody never take no cutoffs and hury along as fast as you can.*

Not all the survivors recommended that anyone hurry along to California, though. Fifteen-year-old Mary Murphy spoke for many of those who had been orphaned by the ordeal when she wrote that same month,

> *i hope i shall not live long for i am tired of this troublesome world and i want to go to my mother.*

By May, Sarah and her siblings had been transported to Sutter's Fort. On May 23, Sarah picked up a pen and wrote a letter to her uncle Jonathan and aunt Nancy Blaisdell in Indiana. She wrote clearly and frankly about some difficult subjects, but with a degree of emotional detachment, and she made clear from the outset that she was not prepared to go into particulars about certain things.

Dear Uncle and Aunt—

It is with a heavy heart that I inform you of our mournful situation. I cannot enter into the details of our sufferings; I can only give a brief account. We got on well to Fort Bridger, there we took Hastings Cut Off and became belated and caught in the California Mountains without any provisions except our worked down cattle and but few of them. . . . They made snowshoes, and on the 16th of December, father, Mary Ann, my husband, myself and eleven others set out. . . . We got lost but resolved to push on, for it was but death any way. . . . It snowed for three days and all this time we were without fire or anything to eat. Father perished in the beginning of this storm, of cold; four of our company died at that place.

As soon as the storm]ceased we took the flesh of the bodies, what we could make do us four days and started. We traveled on six days without finding any relief. On the night of the 6th of January, my husband gave out and could not reach camp. I staid with him, without fire; I had a blanket and wrapped him in it, sat down beside him and he died about midnight, as near as I could tell. . . . On the 18th of January we got relief from the settlement. Seven out of fifteen got in to tell the sufferings of the camp.

Sarah went on to briefly describe her mother's and her brother's deaths at Starved Camp, but the letter, at least as it later appeared in the *Lawrenceburg Western Republican,* then terminated abruptly with a string of asterisks. Whether she had come to the end of her ability to talk in a composed way about what had happened or someone had found the rest of it too personal to print, we do not know.

Mary Ann also wrote a letter, the day before Sarah did. She addressed hers to Levi Fosdick, Sarah's father-in-law, perhaps because Sarah could not bring herself to tell Jay's father what had happened, even in a letter. Like Sarah's, Mary Ann's letter sketched out the sufferings of the snowshoe party and then revealed that Jay had died, "the idol of his loving wife." Unlike Sarah's letter, Mary Ann's was rife with emotion, and it came to a conclusion that differed greatly from Virginia Reed's and revealed much about the surviving Graves children's state of mind as they contemplated how they had arrived in California and what they thought about it.

I have told the bad news, and bad as it is I have told the best. No tongue can exceed in description the reality. . . . I will now give you some good and friendly advice. Stay at home—you are in a good

place, where, if sick, you are not in danger of starv-
ing to death. It is a healthy country here, and when
that is said, all is said. You can live without work if
you are a complete rascal; for a rascal you must be
if you are to stand any chance at all. In the number
of rogues this country exceeds I believe any
other. . . . I have said enough in favor of the
country—as much and perhaps more than I ought.

Mary Ann Graves

Oddly, Mary Ann's letter, at least as published, did
not mention a significant piece of news—that several
days before, she had married a young man named
Edward Pyle Jr.

Pyle had been among a number of men who had
worked to supply logistical support to the relief parties,
ferrying supplies up into the foothills and helping to
escort survivors back down to Johnson's Ranch or on to
Sutter's Fort. One of those whom he had escorted was
thirteen-year-old Virginia Reed. Pyle, like most young
American men in California that spring, knew just how
scarce women were in the West. He had promptly proposed to Virginia and just as promptly been rebuffed.
When he saw Mary Ann Graves, he had tried again,
and this time he'd won.

Mary Ann's quick marriage must have been a great relief for Sarah. It meant one fewer mouth to feed. Living at Sutter's Fort that May, Sarah found herself in a land of extraordinary abundance, with no way to share in it. Green fields of new wheat stretched in all directions around the fort under clear, warm skies. Thousands of head of fat cattle grazed on vast tracks of uncultivated land. Hundreds of Indian vaqueros rode herd on the cattle, tended the fields, and hurried about carrying goods to and from Sutter's Landing on the south side of the American River. Inside the fort's high adobe walls, Sutter's Indian guards drilled with military precision in full uniform on the open parade ground. Blacksmiths, weavers, bakers, and various other sorts of craftsmen worked in shops recessed into the fort's walls. The smell of barbecued beef hung in the air morning, noon, and night.

The emigrants living within the fort, and those camped in a ramshackle cluster of white-topped wagons and makeshift cabins outside, worked and ate and drank and sang lustily, grateful to finally be in California. But among them wandered three little girls, politely asking for food. "We are the children of Mr. and Mrs. George Donner," they said. "And our parents are dead."

· · ·

Sarah and her siblings weren't much better off than the Donner girls. In addition to having little cash and no source of income, they also faced an immediate debt. On April 7, Sutter had asked James Reed to draw up an accounting of what the rescued emigrants owed him for the mules and provisions he had sent over the mountains with Stanton the previous October. The largest amount due on the ledger was the Graves family's share: the hefty sum of $89.93.

Sarah did have one asset, though, at least in principle, and she set out to capitalize on it. At the lake camp, Elizabeth Graves had sold two head of oxen to Margret Reed, to be paid back two for one in California. As the oldest living heir, Sarah calculated that she had four head of cattle, or the money they represented, coming to her, and she wanted one or the other as soon as possible. She also had come into possession of some items that belonged to the Reeds, probably things salvaged from the double cabin, and showing some of the tenacity her mother had shown at the lake, she intended to hold on to them until she was paid for the cattle. In early May she went to the alcalde, John Sinclair, who began negotiations with James Reed by mail.

Reed, though he had presumably arrived in California penniless himself, had gotten off to a surprisingly promising start. He already owned a large swath of land in the Santa Clara Valley, and he was beginning to wheel and deal in real estate and other commercial activities that would in time make him a prosperous and prominent citizen of San Jose. One of his first transactions, in May, was to form a partnership to buy a herd of horses. His partner was another wheeler and dealer, Lansford W. Hastings.

In the meantime Sarah struggled to feed and clothe herself and her surviving siblings. Billy removed himself from the equation when, somewhat improbably, he decided to go home to Illinois and joined an expedition heading east over the Sierra. On June 3, Sarah attended an auction of items recovered from George Donner's wagon and tent. She spent $31.18, money she had likely gotten from the sale of goods salvaged from her own family's side of the double cabin. Virtually all of it went for material with which she could make clothes for her little brother, Jonathan, and her sisters—muslin cloth, woolen fabric, shirting, and some indigo dye. She also bought some black calico with which to make widow's weeds for herself. Other than that, she bought only a few essentials that young women would need—combs and hairbrushes and a spelling book. Someone needed to teach the younger children how to read and write, and for all that she was a sister, Sarah was also now a mother to her youngest siblings.

GOLDEN HILLS, BLACK OAKS

As the wet spring weather of 1847 gave way to the bone-dry heat of a California summer and the hills turned from green to gold, most of the survivors of the Donner Party, and those who had rescued them, spread out over a land that was astonishingly vacant. Most of the Sacramento Valley had never been tilled. The missions along the coast had long since been abandoned—their fertile fields lay fallow, their orchards had gone to weed, their adobe walls had begun to crumble, their darkened chapels stood quietly empty. In the midst of sprawling ranchos of tens of thousands of acres stood only single, lonely adobe homes. In the midst of vast arable plains stood only drowsy pueblos like San Jose and Los Angeles, where mangy dogs sat in the middle of dusty streets, idly scratching their fleas.

Sarah and the rest of her siblings, though, remained in the vicinity of Sutter's Fort that summer. Neither one-year-old Elizabeth nor seven-year-old Jonathan had really recovered from their ordeal, and they had begun to fail. By the end of the summer, both were dead and buried near the fort.

On August 27, Sarah finally received and signed a receipt for the forty dollars James Reed owed her. But for Sarah Fosdick and Lovina, Eleanor, and Nancy Graves, there was really no choice now but to fall back on the charity of others. Over the following months, the four of them lived under different roofs, with different families, first in the vicinity of the fort and then in San Jose, where Mary Ann and her new husband had settled.

By the following spring, though, Sarah and Eleanor had left Lovina and Nancy with the Isaac Branham family in San Jose and moved to the presidio at Sonoma. There they lived for a time with Matthew Ritchie and his family, the first people who had taken her in at Johnson's Ranch. Then one day Reason Tucker showed up and talked Sarah and Eleanor into moving over the hills to the Napa Valley.

It is hard to imagine a more bucolic place than the Napa Valley in the spring of 1848. In the great, broad swale lying between the Mayacamas Mountains and the Vaca

Hills, the soil was so fertile, the climate so beneficent, the living so good that in time this would become some of the most valuable real estate on earth. In the mornings the air was clear and dry and smelled of bay laurel. The sun crept across improbably blue skies, warming green hillsides ablaze with orange poppies. On the parklike savanna of the valley floor, herds of elk grazed under the sprawling arms of valley oaks. Cool streams burbled down the sides of the mountains, emerged from dense, dark stands of ancient redwoods and red-barked madrone trees, then flowed out across the valley and emptied into the Napa River. The river meandered lazily down the center of the valley, meeting new streams, growing fatter and lazier as it went, all the way to the cold, green, saltwater chop of San Pablo Bay. In the afternoons the sunlight slanted in low over the Mayacamas and lit up the peak of stately Mount St. Helena rising above the valley to the northeast, painting it purple and gold.

Up to that time, fewer than a dozen American and Mexican families had shared in the valley's bounty. So when the first few of the new wave of emigrants began to straggle into the valley in 1847 and 1848, they discovered a place that exceeded even the wildest dreams they had harbored on cold, sleety nights back in Illinois or Missouri. Everywhere they looked, they saw opportunities.

Some of the most attractive opportunities they came across lay on a rancho at the far northern end of the valley, where a number of mountain streams provided easy and ample sources of fresh water. The land belonged to an eccentric Englishman named Dr. Edward Turner Bale. Bale had arrived accidentally in California in 1837 when the ship on which he served as surgeon, the *Harriet,* was wrecked off of Monterey. Mariano Vallejo, glad to have such medical expertise wash up on California's shore, had promptly named him surgeon-in-chief of the Mexican forces in California. Bale had wooed and married Vallejo's beautiful niece, María Guadalupe Soberanes, and Vallejo had given the couple the seventeen-thousand-acre rancho to get them off to a good start. Bale, oddly, christened the property with what turned out to be a strangely apt name for some of its subsequent residents, "Rancho Carne Humana"—Human Flesh Ranch.

By the time he saw Americans coming into the valley in the mid-1840s, Bale was rich in property but poor in cash, and he was eager to sell pieces of his property to the new arrivals. Knowing and trusting one another as perhaps no one else after all they had been through, the new settlers were attracted to the idea of banding together as neighbors, so many of them were eager to buy land from Bale. A number were people Sarah had come to know on the overland trip and the subsequent attempts to rescue the Donner Party. The Tuckers, the Kelloggs, and the Starks all purchased adjoining parcels from Bale.

Within a short time, a tight community was born, centered on their mutual interests, their shared history, and a single building. As the settlers began to sow the fields he had sold them, Edward Bale realized that they would need a place to mill the wheat that most of them preferred to grow. So he began construction of a large gristmill at the foot of one of the mountain streams tumbling down the west side of the upper valley. The mill—said to be the largest of its kind in the United States at the time—featured a twenty-foot overshot waterwheel fed by a long redwood flume, and wooden cogs.* When set in motion, it made a thunderous clattering that could be heard for miles. But it was a vast improvement over Bale's previous mill, which had been powered by Indians made to walk in endless circles, rotating the millstones by hand. The new mill was housed in a towering three-story building made of redwood planks cut at Bale's nearby sawmill. The building served primarily as a granary, but in short order it also became a commercial center, a dance hall,

---

* The wheel was later replaced with an even larger, thirty-six-foot wheel. The Old Bale Mill remains a popular visitors' attraction today.

and the community gathering place for the burgeoning American settlement.

The new community was largely self-sufficient, except for one thing. With a sizable population of young children and more on the way, it needed a school, or at least a schoolteacher. Reason Tucker knew that the young widow living in Sonoma, Sarah Fosdick, was lettered, bright, and in desperate need of employment, so Tucker rode to the Ritchies' place in Sonoma and fetched her.

Sarah moved in first with the Tuckers and then with the Kelloggs, who were running the mill for Bale. Someone constructed a brush shanty under a tall fir tree out in front of the mill to serve as an open-air schoolroom. Sarah placed some benches in the shanty, gathered a few schoolbooks, and set about her new job, finally able to earn a bit of money to help support herself and her sisters.

In May of that year, though, just as things were finally beginning to settle down for the Graves girls, Mary Ann's young husband, Edward Pyle, went missing in San Jose. Distraught and fearing the worst, Mary Ann searched through the brush along Almaden Creek near their home, but she could find no sign of her husband. Days, and then weeks, and then months dragged by, and still he did not show up. Mary Ann didn't know if he had fallen ill, or died, or simply left her. She found out only after his bones were discovered.

Edward Pyle had apparently had the misfortune to get into a dispute over a horse with one Antonio Valencia. Valencia had lassoed Pyle and then dragged him behind his horse for a mile. When Valencia finally dismounted and found the young man still alive, he cut his throat. Then an accomplice, Anastasio Chabolla, shot the body full of arrows to make it appear that Indians had committed the murder, and the two rode off. Nearly a year later, after Pyle's remains were found, Valencia was arrested, tried, found guilty, and sentenced to hang. Mary Ann Pyle prepared food for the condemned man and delivered it to him in his jail cell every day, to make sure he lived long enough to hang. On May 10, 1849, he did.

Lovina began to come more frequently to visit Sarah and Eleanor in the Napa Valley, where they lived at different times with the Kelloggs and the Tuckers. The younger two girls waited tables in boarding-houses and took whatever other kinds of work they could find. Meanwhile, they set about the more serious business of finding husbands, the only real path to economic stability open to young single women in the 1840s.

At first the pickings were decidedly slim. Men vastly outnumbered women in California, but in May of 1848 almost all the young, eligible men of the Napa Valley had headed for the hills, looking for more gold of the sort that Peter Wimmer and John Marshall had plucked out of a millrace on the American River in January. By late that summer, though, many of them were beginning to return, and some of them had a good deal of jingle in their pockets.*

After a full year of mourning for Jay, Sarah was looking as well, and by the fall of 1848 she believed she'd found a man who would suit her. She had actually met him long before Jay died, back in another world, when they had all first set out across Missouri. And then much later he'd been among the first people she saw when she staggered half naked out of the mountains at Johnson's Ranch. William Dill Ritchie was Matthew Ritchie's only living son, and likely therefore to inherit much of his father's considerable wealth. Like Edward Pyle, he had helped supply logistical support for the First Relief. He was three years younger

---

* While we generally think of the Gold Rush as having occurred in 1849, for those few who were fortunate enough to already be in California, it began in the spring of 1848. My great-uncle, George Tucker, was among those who were first on the scene, and by the summer of 1848 he had gathered enough gold to buy his own spread on the floor of the Napa Valley, at the age of eighteen.

than Sarah, not much more than a boy, but all in all he seemed like a good choice.* In October of 1848, Sarah married him.

By the following year, Sarah's financial problems and worries were mostly resolved. In September of 1849, Eleanor married William McDonnell, who had been a teamster for the Kelloggs on the overland trip, traveling just ahead of the Graves family on the Hastings Cutoff. The Branham family in San Jose had taken in Nancy more or less permanently. Matthew Ritchie moved his family from Sonoma to the Napa Valley and bought a large parcel of land adjoining the Kellogg, Stark, and Tucker properties. A substantial chunk of the land was for William and Sarah to homestead as their own.

And so they settled down. Sarah continued to teach school out in front of the Bale Mill, and on November 22, 1849, she gave birth to their first son, George Leet. Two years later, in 1851, she and William had another son who died shortly after birth. Then, in 1853, Sarah bore a third son, Alonzo Perry. Living with her children in a small house out in the middle of perhaps the most beautiful valley in the world, Sarah must have felt that she had finally realized at least some

---

* Not everyone thought Ritchie a good catch. Georgia Donner later recalled that some who knew Ritchie thought him to be "unworthy of her." Georgia was only six at the time of the marriage, however, and her sources might have formed that opinion later, and perhaps in light of subsequent events.

measure of the dream she and Jay had nourished lying under the stars back on the plains.

Even in this seeming paradise, there were problems, though. Sarah soon found that her young husband was prone to wander—to take his gun and disappear for days at a time. His frequent sojourns, her sisters reported, often left her feeling unsettled, lonely, and vulnerable in her dark house. Sometimes Eleanor or Lovina came to stay with her for company.

One such night, when William was away, Lovina went outside and found that one of Sarah's young pigs had gone missing. A few moments later, a bear charged out of the darkness, chasing the pig toward the house. She and Sarah screamed and waved their arms and succeeded in distracting the bear for a moment, but then it resumed chasing the pig. The two animals began to run in circles around the house. Sarah and Lovina stood in the doorway, and each time the pig passed by, they ran out and tried to direct it into the house before the bear could catch up. But with each circuit of the loop, the circle widened and the bear began to gain ground. At last it caught the pig and carried it away into the brush, screaming as only a dying pig can scream.

Finally, one day in May 1854, William Ritchie didn't come home at all. At about the same time that he vanished, a pair of mules also disappeared from a ranch in Sonoma County. The owners of the mules—farmers named Tarwater and Hereford—tracked the missing animals over the hills into Napa County, following a trail left by a dangling lariat, but then they lost the trail. They sent a description of the mules to the sheriffs of other counties, and in time they received a letter back from the sheriff of Shasta County that one William Dill Ritchie had been apprehended there trying to sell a pair of mules matching the description. Tarwater and Hereford journeyed to Shasta City and returned with the mules and William Ritchie.

They took Ritchie to the Carrillo Adobe, in present-day Santa Rosa. There a group of men gathered to consider what to do with the young man. Some were for hanging him forthwith. Others argued that the law should take its course. Those with cooler heads prevailed, mostly at the urging of a local farmer named J. E. Davidson. That evening a guard of armed men set out with Ritchie to take him to the county seat at Sonoma to stand trial. Among the guard was Tarwater, one of the aggrieved parties.*

---

* Tarwater was likely Martin Tarwater, a crotchety Sonoma County farmer whom Jack London met many years later. London lampooned Tarwater in a short story called "Like Argus of the Ancient Times." London's fictional "John Tarwater" was hotheaded, oafish, and particularly averse to legal proceedings. "The application of lawyers to John Tarwater was like the application of a mustard plaster," London wrote.

Sometime around midnight that night, the party escorting Ritchie stopped in a grove of oak trees on a ranch belonging to General Joe Hooker, a veteran of the Mexican War who a little less than a decade later would find himself commanding Abraham Lincoln's entire Army of the Potomac. Whether others intervened or the guards themselves took matters into their own hands is unclear, but someone dragged Ritchie from his horse, put a noose around his neck, and threw the other end of the rope over an oak limb. There had been suspicions all along that Ritchie had had an accomplice, and they wanted to know who it was. They hauled Ritchie up into the tree by his neck for a few moments and then let him drop. They loosened the noose and demanded to know who had helped him steal the mules. Ritchie gasped out that he was innocent, that he had won the mules on a bet. They strung him up again, then again let him down and asked him the same question. The answer was still the same, though, so they tightened the noose and pulled William Ritchie up into the air one last time. He kicked and bucked and writhed. His face turned red. His eyes bulged out. His tongue protruded from his mouth and began to turn black. And finally he died, twitching at the end of the rope.

Some days later a coroner's inquest was duly convened. Members of the guard were called as witnesses, but somehow none could quite recall what had happened. All of them, it was later said, wore oak sprigs in their buttonholes.

And at some point someone went to Sarah Ritchie's door at the little house out in the middle of the Napa Valley and told her that once again she was a widow. This time a widow with two toddlers.

16

PEACE

Eighteen fifty-five was a marrying year for the Graves girls. In February, Nancy, now sixteen and an enthusiastic convert to Methodism, married a young Englishman, a Methodist minister named Richard Wesley Williamson, and set out on what would turn out to be a lifelong saga of missionary work. In June, Lovina married John Cyrus, who had traveled the Hastings Cutoff just ahead of the Donner Party. A year or two earlier, smallpox had swept through his household in the upper Napa Valley, killing his father and two brothers and leaving John as the head of the family. And on Christmas Eve that year—a year and a half after Ritchie's lynching—Sarah went to Eleanor's house and stood in the parlor and married for a third time. Her new husband, Samuel Spires, was a farmer, a thirty-seven-year-old widower with two children from a previous marriage. He was by all accounts a good and kindly man, and that, above all, was what Sarah needed.

It's likely that the marriage was more pragmatic than romantic in its origins, as so many nineteenth-century marriages were. At thirty, Sarah probably saw in Sam Spires a ready form of salvation—a means of recapturing for herself and her children some of the economic security that a gang of vengeful men had deprived them of just eighteen months before. And if Sam Spires, a melon farmer, was not an adventurous younger man, so much the better. By now Sarah had likely had her fill of adventurous young men.

Over the next ten years, Sarah and Sam Spires continued to live and farm in the Napa Valley. Sarah produced four more children—Lloyd in 1857, William in 1861, Eleanor in 1862, and finally Alice Barton in 1865. After Alice was born, they packed up their family and left behind the valley in which Sarah had lived for twenty years now. They lived for a time in Visalia, on the eastern side of the San Joaquin Valley, where Mary Ann and a new husband of her own, James Clarke, had settled. Then they moved on to Corralitos, a small community nestled in a valley southeast of Santa Cruz on the coast.

. . .

Just over a range of hills from the wide blue crescent of Monterey Bay, Corralitos was a world unto itself immediately after the Civil War, as it is today. It was only a village—perhaps twenty houses, a half-dozen stores, a flour mill, a blacksmith shop, a tannery, and a schoolhouse nestled at the base of a redwood-covered hill. But it lay in one of the most charming settings in California, a much smaller rival to the beauty of the Napa Valley. At the head of the Corralitos Valley, where the village sat, a creek poured down out of the redwoods, powering both the flour mill and a pair of sawmills higher in the mountains. Wheat fields spread out on the valley floor. In nooks and glens off the main valley, there were new apple, pear, cherry, and plum orchards—fluffy and white with blossoms in the spring, stately and green in the summer, and heavy with red and golden and purple fruit by the approach of fall. To the south the valley opened out into the large, fertile, and almost perpetually foggy plain surrounding Watsonville, where the Pajaro River wandered down to the bay.

Early in the mornings and late in the afternoons, fog filtered through the redwoods above Corralitos, muffling the valley with a soft gray blanket. But by midday the sun burned through the fog and warmed the village and the floor of the valley and the homes that sat tucked among the wheat fields and orchards. It was the kind of country that made a man with an inclination to farm want to thrust his bare hands into the rich, warm, black redwood loam and plant something. And that is what Samuel Spires did, planting and tending melons and strawberries and various truck vegetables while Sarah tended at home to her new brood of young children.

Whatever degree of peace and domestic tranquillity Sarah and Samuel knew in Corralitos did not last long, though. On March 28, 1871, at the age of forty-six, lying at home in an old walnut bed with a high headboard and a wide footboard, Sarah died. Her heart gave out. Years later her daughter Alice remembered that her mother looked peaceful but old far beyond her years as she lay in her deathbed. Samuel did not cry when his wife died, but he appeared stunned, much as when, Alice remembered, he had once awakened and found that an early frost had ruined his melon crop just as it came to perfection. When it was over, the house was still. Outside, all across the valley, with every passing breeze, showers of white petals fell silently from apple trees, like snowflakes.

IN THE YEARS BEYOND

In the months and years following their rescue from the mountains, the survivors of the Donner Party went on to live or die as best they could.

Billy Graves returned from Illinois in 1849, guiding a party of Gold Rushers. When the company reached Donner Lake, Graves disappeared for an afternoon. He might simply have gone hunting, he might not have wanted his fellow travelers to see his distress at revisiting the site of his suffering, or—and this is purely my own speculation—he might have gone searching for the cache of coins his mother had buried three years before. He settled in Calistoga, just north of the little community surrounding the Bale Mill where his sisters lived, and took up blacksmithing and prospecting. He married a Pomo Indian woman, then later abandoned her, and in 1873 married Martha Cyphers, from whom he was later divorced. In his old age, he was a notoriously eccentric figure around Calistoga. He died in Santa Rosa in 1907.

Mary Ann and James Clarke lived out their lives on their ranch in Visalia. Mary Ann gave birth to seven children and took in Sarah's young daughters, Eleanor and Alice, for a time following their mother's death. Mary Ann remained a strong and somewhat formidable figure in later years, given to smoking a clay pipe. But she remained emotionally scarred by her Donner Party experiences for the rest of her life. Mary Ann died in Visalia in 1891 and is buried there.

Eleanor Graves and her husband, William McDonnell, acquired a sprawling ranch in Knights Valley, north of Calistoga, where they prospered, and Eleanor bore them ten children, though four died young. At the time they settled on their ranch, it was said to be the northernmost American homestead on the Pacific coast until one reached Oregon. Like Mary Ann, the McDonnells took in Sarah's daughters for a time. Eleanor died at the ranch in 1894.

Lovina and John Cyrus produced six children and lived well in the upper Napa Valley, where they became well-known and beloved figures. Lovina died in 1906 and was buried in Calistoga.

Nancy and her husband moved from town to town in California, spreading the Methodist faith and rearing nine children. Late in life they finally came to roost in Sebastopol, California, where Nancy died in 1907. To the end of her days, she remained averse to talking about the Donner Party ordeal and outright refused to provide information about it to Charles McGlashan.

On May 14, 1891, a prospector named Edward Reynolds was scratching around on a hillside above the western end of Donner Lake when he found a few old silver dollars lying near the rotten stump of a pine tree, about thirty feet from a large boulder. Digging deeper, he found more silver coins. The next day he returned to the site with a compatriot, Amos Lane, and with the editor of the *Truckee Republican*, Charles McGlashan, who had by then become an authority on the Donner Party. The men did some more excavating and soon had a hatful of silver and gold coins, the earliest of which was dated 1845. When he was later shown the coins, Billy Graves identified them as those that his mother had buried in 1847, noting tooth marks that one of the Graves children had left on one when teething as a baby. Half of the coins were given to Reynolds and Lane; the others were distributed among members of the Graves family, whose descendants treasure them to this day.

Peggy and Patrick Breen's family survived the disaster intact. In September of 1847, they left Sutter's Fort and moved to San Juan Bautista, where they lived rent-free in the adobe of General José Castro, who had commanded the Mexican army in Northern California during the Mexican War. In 1849 the Breens' oldest son, John, then seventeen, went to the goldfields and returned with twelve thousand dollars. The Breens used the money to buy the Castro adobe and to acquire large amounts of land in what is now San Benito County. From that point forward, they prospered and became prominent members of their new community, not far from where Sarah died in 1871. Patrick Breen died in 1868, and Peggy Breen followed him in 1874. In 1878, James Breen, just five at the time of the disaster, walked into the offices of the *Truckee Republican* and asked Charles McGlashan if he could subscribe to the paper. When McGlashan learned that the young man was a survivor of the Donner Party, he became so intrigued that he launched an intensive effort to research the story. His effort culminated six months later in the first full-fledged book on the subject, *History of the Donner Party*.

James Reed's family also emerged intact, and they also prospered. As a result of real-estate transactions and mining successes, Reed acquired both wealth and social prominence in San Jose, presiding over large tracts of land, including the one on which San Jose State University now stands. Reed continued to have what some considered an outsize opinion of himself. In July 1847, he wrote home, "Our misfortunes were the result of bad management. Had I remained with the company, I would have had the whole of them over the mountains before the snow would have caught them; and those who have got through have admitted this to be true." Margret Reed remained frail after the disaster and died in 1861; James died in 1874. Virginia Reed converted to Catholicism after witnessing the strength that the Breens' faith had given them in their cabin at the lake. She married and bore nine children and died in 1921. Patty Reed became one of the public faces of the Donner Party in her later life, partly because of the endearing story of the small wooden doll that she had carried across the plains and kept hidden at the lake camp, but mostly because of her ceaseless efforts to keep alive the memory of those who had suffered. She bore eight children and was widowed at a relatively young age. She supported her children by keeping a boardinghouse in Santa Cruz and died there in 1923.

After they left Sutter's Fort, George and Tamzene Donner's orphaned daughters were taken in by other families, including the Reeds, who took in Frances. All the girls married, lived full lives, and bore children. Georgia died in 1911, Frances in 1921, Eliza in 1922, Elitha in 1923, and Leanna in 1930.

Three children from Jacob and Elizabeth Donner's family survived as orphans. Ten-year-old George Donner was first taken to San Francisco, where he lived rent-free in a hotel and where Lansford Hastings provided him with some of his clothes. He later became a farmer, but he died in 1874, just thirty-seven years old. Several of Mary Donner's injured toes were removed at Johnson's Ranch immediately after her rescue. She, like her cousin Frances, was taken in and raised by the Reed family in San Jose. She married in 1859 but died within a year, apparently from complications arising from childbirth. Solomon Hook worked variously as a carpenter, an innkeeper, and a farmer and died of cancer of the jaw in 1878.

William Eddy, having lost his wife, Eleanor, and his two children in the disaster, moved to the San Jose area. In 1848 he remarried, and he and his wife had three children, including one whose name stood as a testament to Eddy's political views—James Knox Polk Eddy. He subsequently divorced and remarried again and died in Petaluma, California, on Christmas Eve 1859.

Amanda and William McCutchen, mourning their baby, Harriet, moved first to Sonoma and then to

Gilroy in the Santa Clara Valley, where William was elected sheriff in 1853. In 1857, Amanda died in childbirth. William remarried and lived until 1895, when he died of a stroke.

In June 1847 the young widow Harriet Pike married Michael Nye at Sutter's Fort. The couple raised stock in the Marysville area for a few years and then moved to The Dalles in northern Oregon, where Harriet died in 1870. Her daughter Naomi, brought out of the mountains by the First Relief at the age of two, married a banker and became his wealthy widow, until the stock-market crash of 1929 wiped out her fortune. She died in 1934.

Mary Murphy, fifteen and an orphan when she was brought to Johnson's Ranch by the First Relief, married the ranch's proprietor, William Johnson, that same June. Johnson soon proved to be an abusive husband and, in Mary's own words, "a drunken sot." Mary divorced him. On Christmas Day in 1848, at Sutter's Fort, she married again, this time a man named Charles Covillaud, fifteen years her senior. The couple prospered through the Gold Rush and produced five children. The town that grew up where they settled at the confluence of the Yuba and Feather rivers was named Marysville in her honor. But Mary's husband died in February of 1867, and Mary followed him in September of that same year, dying at the age of thirty-seven.

Mary's brother William became a lawyer and a prominent citizen of Marysville and died there in 1904. Her youngest brother, Simon, served in the Union cavalry in the Civil War and died in Tennessee in 1873.

Of the thirteen single men who entered the Sierra Nevada with the Donner Party, just two—Noah James and Jean Baptiste Trudeau—left the mountains alive. Both had been only about sixteen at the time of the tragedy, and their youth likely contributed to their survival. In 1847, Trudeau reported to a naval officer, Lieutenant H. A. Wise, that before he had left the Alder Creek camp, he had "ate Jake Donner and the baby [Sammy Donner], 'eat baby raw, stewed some of Jake, and roasted his head, not good meat, taste like sheep with the rot, but sir, very hungry, eat anything.'" Trudeau spent much of his life as a fisherman on Tomales Bay in Marin County. Later in life he styled himself as the Donner Party's guide. He died in 1910. Noah James disappeared into California in the spring of 1847 and was not heard of again until 1851, when a horse thief with the alias of "Mountain Jim," but reputedly Noah James in fact, was hanged.

Louis and Philippine Keseberg were reunited when Louis limped down out of the mountains with the Fourth Relief. Keseberg worked for a time for John Sutter. He then suffered a series of business setbacks, first buying a hotel that burned down, then operating a brewery that was flooded out in Sacramento. After working at a distillery in Calistoga, Keseberg finally returned with Philippine to Sacramento. Philippine— after bearing Keseberg eight consecutive daughters— died in 1877. Keseberg—publicly and privately reviled as a voluntary cannibal and likely a murderer as well from the time he arrived in California—watched his wife and all but one of his daughters predecease him before he died penniless and friendless in a charity hospital in 1895.

The Reeds' family cook, Eliza Williams, hiked out of the mountains with the First Relief. She married a German emigrant named Thomas Follmer at Mission San Jose in September 1847. After a brief period in Sonoma, she lived most of her life near the Reed family in San Jose. Doris Wolfinger also hiked out with the First Relief and married George Zinz, an Alsatian emigrant, that same year. The couple settled first in Sacramento and then moved to Sutter County, where she died in 1861.

Isabel Breen, the last survivor of the Donner Party— only about one year old in 1846—died on March 25, 1935, in Hollister, California, at the age of eighty-nine.

Lansford W. Hastings lived an odd and peripatetic life following the disaster. After briefly serving in the Mexican War, he set himself up as a lawyer in San Francisco, abandoned his practice to rush to the hills in search of gold, and then moved to Sacramento to go into business with John Sutter. When he had a trunk shipped from San Francisco to Sacramento, it went missing, along with fifteen hundred dollars in gold coins it contained. He nevertheless opened a retail store with Sutter, but the business promptly collapsed, causing Sutter to remark, "The store made money, but I lost. Hastings was a bad man." Bad man or not, Hastings was appointed judge of the northern district of California and represented Sacramento at the state's constitutional convention in 1849. He went on to suffer an additional series of business failures, became the postmaster of Fort Yuma, Arizona, and then hatched a wild scheme to use the Colorado River as a means to channel goods from San Francisco to Brigham Young's burgeoning Mormon community on the shores of the Great Salt Lake. The scheme was designed, not incidentally, to create a financial empire for himself in the Southwest. He was entirely undeterred by the fact that no one had any idea whether the Colorado River was navigable through the Grand Canyon, as in fact it was most emphatically not. Hastings then served as a major in the Confederate army during the Civil War, in which capacity he promulgated a scheme for conquering the

Southwest for the Confederacy. Finally, after the war, embittered by defeat but undiminished in his capacity for pursuing bad ideas, he set out to create a Confederate colony in Brazil and sat down to write *The Emigrants' Guide to Brazil*. He died on a ship returning to Brazil sometime thereafter.

EPILOGUE

On a Sunday morning in March of 2008, I drove down the Pajaro River just west of Monterey Bay, traveling among fields of gray-green artichokes in a dense fog rank with the reek of salt and seaweed. I pulled off the highway and wandered through the derelict streets of downtown Watsonville until I came across the old Pioneer Cemetery on Freedom Boulevard. There I got out of the car and started to look for Sarah's bones.

For more than two hours, I slowly walked the dank cemetery, approaching it row by row, peering through the fog at each and every stone that seemed weathered enough to date back to 1871. Maybe two dozen times, my heart leaped to my throat when I found old marble stones inscribed SARAH, but it was always the wrong

Sarah. Surely there was no more common female name in the nineteenth century.

Finally I came to the last stone of the last row, still with no results. I had seen a number of plots with no markers or with wooden markers from which the names had long since faded away, though. And Sarah may lie in one of them. The cemetery in Watsonville was only about a forty-minute wagon ride from Corralitos in 1871, and it may not be remarkable that her grave was never marked, or marked with something less durable than stone. Though many of the Donner Party survivors were well known as such and much was made of their deaths and burials, there seems to have been little to no awareness in her own community of Sarah's role in history. Two days after she died, the local paper, the *Pajaronian,* noted simply that "a Mrs. Spires of Corralitos fell dead on Tuesday last. Heart disease was the cause." So the anonymous Mrs. Spires's bones may in fact occupy an unmarked, fog-shrouded grave in Watsonville. But I don't think so.

I returned to my car and drove north on Freedom Boulevard toward Corralitos. The sun had begun to burn through the fog by now, and the valley opened up before me in warm, yellow light—a bouquet of vivid green fields, bright yellow mustard flowers, and fruit trees draped in veils of white blossoms. By the time

I reached the village, the sky overhead was blue porcelain and the last few wisps of gray fog were just melting from the tops of the redwoods up in the hills.

I drove to a certain address on a certain street at the western edge of town, got out of the car again, and stared up at a hillside cloaked with a dense tangle of madrones, oaks, bay laurels, and acacias dripping long tassels of bright yellow flowers. The hillside, a local historian had told me, was once home to the Corralitos Cemetery, back in the 1870s. But in 1878 the owner of the property decided he wanted to put in an orchard. He notified the local families that they needed to remove the remains of their loved ones. By then, though, Samuel Spires and his children had moved away from the valley.

Many of the bodies were duly exhumed and moved to Watsonville or elsewhere, but others were still in place when a hired man began to prepare the ground for the orchard. Careless or overzealous with his plow, he destroyed the markers and obliterated any signs of the remaining graves. The man was fined, the orchard was planted, and the dead have remained in place ever since.

Peering into the woods that have since replaced the orchard, it occurred to me that this bit of jungle was much like the tangle of trees through which a barefoot six-year-old girl named Sarah Graves once wandered in the bottomlands along the Illinois River, looking for secret treasures. I pondered whether this was indeed her final resting place. I could not be sure, of course, but standing there with the warm California sun on my back, awash in the delicious spice of bay laurel, listening to the jabbering of a tribe of scrub jays in the oak trees, I decided to hope so and, finally, to believe so. The place was wondrously alive, joyously fecund, welling up with something intangible but vibrant. It was, at the very least, where Sarah ought to be.

My visit to Corralitos ended the odyssey I had begun when I visited my great-uncle George Tucker's bones in the Napa Valley in the fall of 2006. In the interval I had traveled some sixteen hundred miles in Sarah's footsteps, trying to get a sense of what she had seen and felt in 1846 and 1847.

Along the way the landscape of the American Midwest and West had offered up many details of touch, taste, aroma, sight, and sound that Sarah and her companions never recorded and that I would never have been aware of without traveling the route. Beyond the sensory details, though, my travels offered me a context in which to muse on Sarah's story, on what to make of so much suffering and so much courage. And so as

I traveled, I tried not just to breathe in the dust and feel the glare of the sun but to listen to the quiet whisperings of the bones I passed along the way—bones long lost in tall, blue-green prairie grass, bones scattered across the alkali deserts of Nevada, bones now turned to rich black humus in quiet mountain meadows—to hear what they had to say about Sarah and her companions and to ponder what it all meant.

I began in Sparland, Illinois—the meager village that has grown up on the land that Franklin Graves sold to George Sparr in 1846. I drove into town early on the morning of April 12, 2007, exactly 161 years to the day after Sarah had left.

I made my way to the top of the limestone bluff that rises above the river to get a view of the landscape. When I stepped out of the car to take a few pictures, a frigid west wind sliced through my many layers of clothing, numbed my fingers, clawed at my face, brought tears to my eyes. I hastily snapped a few photos and dove back into the shelter of the car.

From within the warm cocoon of the car, I surveyed the place where Sarah grew up. A couple of dozen small white houses clung to the side of the bluff below me. At the base of the bluff was a crossroads with a gas station, a grubby convenience store, and a small brick post office. Nearby stood an old frame house on the site where the Graves cabin once stood—far enough back from the river to avoid flooding but close enough to allow access to the water. To the east of that, the land sloped gently down into the muddy fields where Franklin Graves once grew his wheat and corn, and beyond that to a swath of deciduous woods fringing the river. The river itself, dark and overflowing its banks as it was the day that Sarah left, was spanned by a steel-and-concrete bridge—an edifice that would have astounded Elizabeth Graves could she have looked up and contemplated it on one of those winter days when she rowed across the river to deliver herbs and advice to her neighbors.

Looking at the bridge, I wondered if I would be able to bridge the distance to Sarah, to find a common strand of humanity that would enable me to comprehend her world and her travails. To really understand her story, I knew I would have to travel farther than just the sixteen hundred miles that lay between me and California. I would have to travel into the heart of a girl who was a product of a vanished world. And not only that, but a girl who encountered in her life challenges more daunting and tragedies more profound than I have ever begun to confront in my own.

A bit ill at ease, I drove down off the bluff, turned left onto the highway, and set out west, starting to follow Sarah.

. . .

In one long day of driving, I covered the ground that Sarah covered in a month—from Sparland to St. Joe, Missouri. As I drove into the old river city that evening, the crumbling brick warehouses and limestone office buildings of downtown, softly aglow in the colors of faded roses and aged ivory, looked something like the ruins of a lost civilization.

I pulled off the freeway and parked in a gravel parking lot and looked west across the Missouri River in the twilight. The broad river slid by, silent except for the slapping of little waves on the concrete riprap under my feet. As I took it in, it seemed to me that the view across the river was the single most compelling thing about St. Joe. That was a kind of paradox, for I could see almost nothing on the far side of the river, only what Sarah must have seen in 1846—a dark line of trees etched against a soft evening sky. I couldn't see anything beyond the trees, and that was what interested me.

For Sarah, everything on the far shore, everything unseen beyond that line of trees, was outside the United States and outside the normal scope of her life. It was a vast unknown, a blank slate on which her entire future and much of her country's future, both real and mythical, were about to be written. Everything she hoped for, and everything she feared, lay beyond those trees, and she could not yet know in what proportions they would be mixed. Whether she and Jay would prosper, what kinds of lives they would live, what sorts of children they might raise, what nation's flag they would live under, what hardships they might be forced to endure, what friends they might make, what deaths they would eventually suffer—all that and more waited, unrevealed, beyond those trees.

Standing there, listening to the Missouri lap at the rocks below me, watching the trees on the opposite bank dissolve into darkness, I wondered how she did it. What species of hope allowed or compelled Sarah and her family to make the leap, to cross the river and venture beyond the trees into so vast an unknown when so many more chose to stay home?

I am, coincidentally, the father of two daughters, one exactly the age that Sarah was in 1846, the other almost exactly the age that Mary Ann Graves was. And I am myself almost the same age that Franklin Ward Graves was in the spring of 1846. I love my daughters beyond words, and I believe that each of them, in her own, quiet way, is courageous. But I cannot imagine either of them undertaking what Sarah undertook. Like Franklin Graves, I understand how it is to see better opportunities elsewhere and to pick up a family and move in order to capitalize on them. But the choices

I have made and the chances I have taken shrink to insignificance compared with Franklin Graves's choices and chances. I cannot begin to conceive of hazarding what he did.

To be sure, he and his family were all no doubt caught up in the great national passion of their time—to conquer the West, to do God's work by seizing what was manifestly their destiny. But in the end I suspected that making the decision to go depended on something deeper and more personal than any of that. They must each have found something in their hearts—some quality of faith and hope so powerful and reassuring that it caused them not to shrink from but to embrace the human heart's seemingly inexhaustible proclivity to populate the future with what it hopes to find there. To say, when asked, "Yes, I will go."

Late in June, I rejoined Sarah's trail near where I had left it on the Missouri near St. Joe. I drove northwest, bouncing along gravel roads through great tracks of the shoulder-high corn that has replaced the sweeping prairie Sarah knew, following the Little Blue River as nearly as I could north to I-80, then hopping onto the freeway and heading west along the Platte River, paralleling the old California Trail that Sarah had followed up the south bank.

Over the next several days, I beat my way westward on I-80, taking photos and making notes, stopping en route to get a feel for the land, exploring the south bank of the Platte where Tamzene Donner botanized and Edward Trimble died, then cutting northwest across the dusty, dry hills between the South Fork of the Platte and the North Fork to Ash Hollow, where I hiked among scrubby junipers in the surrounding hills, worrying about rattlesnakes and ticks with every step. I traveled up the North Fork past Chimney Rock and Scotts Bluff and arrived at Fort Laramie early on the morning of the Fourth of July.

I stayed for the day to celebrate the holiday, as Sarah had in 1846, and found that in many ways things had not changed much since she left. The old fort itself was gone, reduced to a crumbling rock foundation and replaced by white wooden structures built later in the nineteenth century. But the spirit of the place seemed to be intact. Hundreds of people had shown up for an old-fashioned celebration put on by the fort's staff and volunteers. There were encampments of RVs and trailers in the shady cottonwood groves down along the river where Sarah and her family had set up their own encampments. Smoke rose from dozens of barbecues; men and little boys walked about the grounds in cavalry uniforms; women and little girls wearing sunbonnets sat in the shade drinking root beer and calling it sarsaparilla. I sat in the shade, too, for much of the day, watching the festivities unfold, leaning against a cottonwood that was perhaps descended from one under which Sarah had sat a century and a half before.

That afternoon, as the sun began to sink toward the rolling, pale gold hills that lay just to the west, I watched the last of the celebrants pack up their cars and RVs, getting ready to drive back home for their evening fireworks. It was an easy leap to conjure up Sarah again, to watch as she packed her wagon down among these cottonwoods on that long-ago Glorious Fourth of July, still wondering, as she must have been, what lay over those golden hills.

In early August, I drove over the South Pass and down the Big Sandy and the Blacks Fork of the Green River to Fort Bridger, now a Wyoming state park, complete with a convincingly shabby replica of Jim Bridger's original trading post. From there I followed I-80 west beneath the spectacular red-rock cliffs of Echo Canyon to Henefer, Utah, where the Donner Party found the note from Lansford Hastings stuck in a clump of sagebrush. Just outside Henefer, I left the interstate and turned up into the mouth of present-day East Canyon.

A pair of sandhill cranes heading south cranked their way laboriously across the sky ahead of me. I made my own way slowly up Highway 65, which approximates the course that James Reed blazed into the Wasatch in 1846.

Winding up the tortuous route, it didn't take me long to see how discouraging it must have been for Sarah and her family. Every canyon the road turned up seemed a little steeper than the one before, a little more densely choked with vegetation. Finally my car struggled up a series of switchbacks to the windswept pass that Reed named "Reed's Gap," though no one since Reed seems ever to have called it that. From a parking lot at the pass, I took in the panoramic vista of the Salt Lake Valley to the west. Trees blocked my view to the east, though, so I decided to hike farther up toward the summit of the peak to my south.

The trail was steeper than I expected. Within five minutes my heart was thudding under my rib cage. The altitude was more than seventy-five hundred feet here. The air was thinner than I was used to, and I was quickly gaining a deep appreciation for how fit Franklin Graves, one year my senior in 1846, must have been when he fought his way over the whole of this range, not just the mile or so that I was attempting. A turkey vulture was circling lazily above the mountain, and I could not shake the notion that perhaps he made it his business to know a fool when he saw one.

Finally, after another thirty minutes of climbing and gasping, I fought through some brush to a spot where the ridgeline fell away to the east. From somewhere very near there, Lansford Hastings had pointed out to James Reed his "better route" through the Wasatch. Looking out at the confusion of green mountains and purple canyons below me, it struck me with full force—in a way that it could not have if I hadn't seen it for myself—that only a madman, or a serious salesman, could look at that landscape and propose taking a party of heavily laden wagons through it.

In mid-September, I flew to Salt Lake City and set out again on I-80. The Great Salt Lake, pale blue and frothy, its verges white with wind-whipped foam and a crust of salt, stretched away to the north. As the interstate led me out onto the salt desert, I noticed that for mile after mile alongside the road, passersby had arranged dark stones into signs, symbols, and messages on the pure white fields of salt. There were hundreds of these stony graffiti—smiley faces, hearts with initials inside, peace signs, crosses, and an occasional odd inscription like POLYGAMY HUMPS.

Most of the offerings, though, were simply initials with dates, manifestations of the ancient urge to record one's passage through an interesting landscape, to leave one's mark. The Donner Party and the other emigrants of their time felt the same urge. James Reed carved his initials, "JFR," and the date, "26 May 1846," on a rock at Alcove Spring in Kansas, where they can still be seen today. And hundreds of Sarah's fellow emigrants in the 1840s etched their names into Independence Rock near Casper, Wyoming, as they passed by it on or near the Fourth of July each year.

Farther up the road, I parked at a rest stop and struck out on foot across the salt. I wanted to see if I could intersect Sarah's trail, to see if it was still visible. Millions of salt crystals glittered and shimmered and crunched beneath my feet as I walked. But it was hot, and after about half a mile the glare of the sunlight off the salt had narrowed my eyes down to sweaty slits. A quarter of a mile farther on, I'd had enough—my eyes were painfully dry, my head was throbbing, and I felt vaguely nauseated. By the time I got back to the rest stop, everything around me seemed to be growing dim and vaguely blackened, like a shadow world. I started up the car, turned on the A/C, put my face to one of the vents, and eagerly sucked up the cool air. My God, I thought, those people were tough.

. . .

Over the next few days, taking my time, I traveled west across the basins and ranges of Nevada—past a spot called Flowery Springs, where the women of the Donner Party all got angry at the same time; down the lazy Humboldt River; past the sandy hills where James Reed killed John Snyder; past the marshes where the Humboldt sinks into the sand and disappears; across the Forty Mile Desert, where plumes of white steam still rise from boiling-hot springs; to Wadsworth, Nevada, where Sarah first encountered the beautifully clear and bracingly cold waters of the Truckee River.

Then I made my way up the Truckee River Canyon on I-80 to Reno. I slid into town on the freeway, exited on Virginia Street, parked behind a casino, and walked through a back door into the cool, clattering, clanging darkness. At row after row of silver slot machines, solemn-looking men and women sat smoking cigarettes, dropping nickels and quarters into the slots, and pulling levers. I sat on a stool with a roll of nickels and watched them play. They worked at their chance taking with a kind of hypnotic rhythm, inserting coins, leaning forward, pulling the lever, settling back to watch the wheel, then leaning forward to insert more coins and pull the lever again.

I thought once more about Sarah and Jay. We all play on a field of chance every day of our lives. But Sarah and Jay played on a particularly dicey field, a particularly deadly one, though they might not have fully understood that. They and their companions had taken an enormous risk simply by selling their farms and businesses and crossing the Missouri. As they had moved west, the stakes had grown higher each time they cast the dice, deciding what to keep and what to throw away, taking the unproven road south at the Parting of the Ways, following Lansford Hastings into the Wasatch and across the salt flats. And then here, just a hundred miles from their goal, where Reno and its palaces of chance would someday stand, some of them had made one final cast of the dice, pausing to rest and regroup in Truckee Meadows before assaulting the Sierra Nevada, even though they could see the snow already accumulating on the peaks ahead of them.

Sitting in the casino, I wondered if the habit of taking chances and thus far surviving them had lulled them into a false sense of security, left them as mesmerized by the temptations of fortune and the hazards of chance as those sitting around me seemed to be as they watched the wheels spin before their faces.

I returned to Reno in late November and drove up into the Sierra Nevada under gray skies. I pulled off the freeway into the town of Truckee, made my way around to the back of the high school, and climbed up on a berm of earth dusted with snow. In front of me lay the westbound lanes of I-80 and a wide spot in the road where, until recently, the old California Agricultural Inspection Station used to stand.

The interstate and the inspection station were built in the 1960s, directly atop the spot where Franklin Graves had built his cabin in 1846. Their construction obliterated any traces of the cabin and precluded any hope of conducting archaeology on the site, but archaeology of a sort *was* conducted in 1879 when Charles McGlashan, Billy Graves, and a number of townspeople from Truckee took picks and spades to the site and began to dig.

By then, people had been sifting through the remains of the double cabin and carrying off relics, perhaps including human remains, for more than thirty years. All the same, McGlashan and the others managed to find a century-old brass pistol, a flintlock rifle, bullets and lead shot, a cooper's in-shave that had belonged to Franklin Graves, and a sealed tin box in which Elizabeth Graves had kept oil of hemlock.

Standing there by the interstate, I found it hard to see past the present, to imagine the outlines of a simple cabin where so much modernity was whizzing by at seventy miles per hour. It was hard to see Sarah and Jay there, stooping over in the snow, putting on their snowshoes, about to begin their trek toward death or salvation. Hard to imagine Elizabeth Graves and Margret Reed standing in the snow, perhaps red-faced with rage, fighting over the hides draped on the roof. Imagination can only take you so far out of your own world. But it occurred to me that any one of the sixteen-wheelers racing by on the interstate could have carried all of the Donner Party over the crest of the mountains in about seven minutes.

I returned to the car, drove across an overpass to Donner Memorial State Park, and parked near the tall monument to the Donner Party.* After taking a photo of the monument, I went into the Emigrant Trail Museum, where a number of Donner Party artifacts were on display, including some of Elizabeth Graves's silver coins. Hung high on a wall in one corner, a picture

---

* The monument is, among other things, a grave marker. On June 22, 1847, General Stephen Kearny, leading an expedition eastward, paused at the lake camp. He ordered his men to gather the human remains and bits of shredded clothing that were scattered about the site and to lay them in a pit they had dug in the floor of the Breen cabin. Then they set the cabin afire and departed. Seventy-one years later, in June of 1918, three elderly ladies—Martha "Patty" Reed Lewis, Eliza Donner Houghton, and Frances Donner Wilder—looked on as the monument, standing on the site of the Breen cabin and the cremated remains, was formally dedicated.

of Sarah—the same image I had been carrying in my pocket for more than a year now—gazed down at me as I worked my way through the museum. Looking up at her, I wondered, not for the first time, if *she* had in fact been looking down on *me* for some time now, wondering what I was up to.

In early January 2008, I returned to the Sierra Nevada. A major winter storm had just blown through, and by the time I got to Truckee, the place looked like Antarctica in July. The streets ran though deep blue canyons of snow. White cornices of snow crowned every building in town, heaped up and sculpted into improbable shapes by the storm's high winds. Shimmering silver daggers of icicles, three or four feet long, hung from every projecting eave.

I drove around the north side of Donner Lake and began to wind my way up old U.S. 40 toward Donner Pass. The road was freshly plowed but serpentine and narrow. As I climbed higher, abrupt cliffs of snow rose on the right side of the road, encroaching on the pavement and crowding me over into the left lane in places; on the left side, only a thin guardrail separated me from the void that fell abruptly away to the lake below. But the road followed the approximate route that the snowshoe party took, and I wanted to get as close as I could.

Finally I parked at an observation point and looked out toward the east.

I was frankly stunned by the beauty of the place— the blue lake below me was just turning to violet in the early-evening light; the snowy peaks surrounding it were tinted gold and pink in alpenglow. Taking in the view, I recalled how Mary Ann Graves had stood near this same spot and, even though she was embarked on a life-and-death endeavor, paused to marvel at the sight of so much grandeur encapsulated in one vista.

The first time I read that she had taken the time to appreciate the view—to let her fancy wander to the image of a troop of Norwegian fur trappers roaming among icebergs—I wondered at the credibility of her report. And I wondered also about its implications. As it turns out, those implications might have been profound, at least for her.

Survival psychologists have since discovered that the people who are most likely to live through extreme, life-and-death challenges are those who open their eyes to the wonders of the world around them, even as their own lives hang in the balance. To appreciate beauty is to experience humility—to recognize that something larger and more powerful than oneself is at work in the environment. And humility, it turns out, is key to recognizing that in order to survive, you must adapt yourself to the environment, that *it* won't adapt to *your* needs. So it seems that Mary Ann Graves carried an advantage with her as she crossed Donner Pass—her attitude. She kept her eyes open; she did not deceive herself. She saw and touched, tasted, smelled, and heard everything that was happening around her, and everything that might.

When I drove down out of the Sierra Nevada on I-80 the next day, I stopped at another scenic overlook, at Emigrant Gap. The freeway there runs along the top of the low ridge that screened Sarah and her companions from a view of Bear Valley and the emigrant road that would have brought them safely to Johnson's Ranch. Getting out of the car to stretch my legs, I could see in a single glance the crux of all that had gone wrong for them. On one side of the road, the ridge falls abruptly and dramatically away to Bear Valley, some seven hundred feet below. On the other side, it falls away more gradually into the wild canyonlands of the American River. This was where they had gone wrong, made their fateful wrong turn. They had been this close to their likely salvation but failed to see it.

In many ways that low ridge seemed to me to be a metaphor for the larger tragedy of the Donner Party, and studying it solidified in my mind a theme that seemed to keep coming back to me wherever I went in their footsteps. From the time they had first encountered Wales Bonney carrying a note from Lansford Hastings back on the approaches to the South Pass, a ridge of deception had slowly arisen between them and the truth of their situation. Led into the wilderness by a lie, led astray at times by their own dreams and ambitions, dazzled by the glare of sun off salt, and confounded by snowstorms, they had found themselves blundering ever more blindly through terra incognita as they moved west. Here at Emigrant Gap, even the landscape itself had conspired to deceive them. And when the land they encountered did not conform to their expectations, they had continued to move forward as if it did, taking the easier route downhill. In the end, as a group they had exhibited precisely the opposite kind of behavior from the humility and open-eyed awareness that survivors always seem to demonstrate.

I left the interstate and drove down into Bear Valley, then down Highway 20 through the foothills to Grass Valley and on into the rolling country where Sarah finally arrived at Johnson's Ranch. I pulled over on the south side of the Bear River. Johnson's adobe, the Ritchies' cabin, and the other ranch buildings that had greeted Sarah had stood fifty yards or so to the north of the spot. It had begun to rain hard, and the river was turgid. The oaks on the other side of the river were stark, black, and leafless.

I stared across the river and recalled the night that Sarah arrived under oaks like these, and the light and warmth she found waiting within the Ritchies' cabin. Despite her bereavement and after the horror of what she had just been through, I suspect there must have been, for Sarah, a moment of utter and absolute relief such as few of us ever know. When she walked or was carried into that cabin, she was finally able to put down a burden that far exceeded any other that she had carried across the Sierra on her back. For thirty-two days, she and her sister had borne the certain knowledge that if someone did not get through to California with word of what was unfolding in the mountains, their mother and all their younger brothers and sisters would almost certainly die. She had survived, and because of that her family still had at least a chance of doing the same. If she wept that night for her father and Jay, it seemed to me that Sarah must also have wept, at least a bit, for joy.

Later that spring I went to Corralitos, to search for Sarah's bones. And then I went home and tried to decide: What—after all my reading and traveling— was I to make of the hard life of Sarah Graves Fosdick Ritchie Spires?

Some have argued that the survivors of the Donner Party were not heroes, that they simply did as anyone would do—fight as hard and as long as they could to survive. On one level I think this is true. Much of the recent research into survivor psychology reveals that what people ordinarily do under extreme conditions is fairly predictable. For the most part, when we are severely stressed, like caged lab rats we bite and claw and squeal until we escape or die. Seen thus, Sarah was simply one of those who struggled and was strong enough to escape.

But I think there is another level to Sarah's story. When I think about Sarah, I think about the nights when Jay serenaded her with his fiddle under a silver spangle of stars on the prairie, and about the night she knelt beside him as he lay dying in the muddy snow of the California they had dreamed about together. I think about the moment when she walked away from her mother for the last time on snowshoes, and about the Christmas Eve she sat shivering by her father as he begged her to use his body for food. I think about every excruciating step she took through the Sierra snow, trying to bring relief to her mother and her siblings. I think about the birth of her first child, and the day men came to tell her that his father had been hanged as a thief. I think about the loveliness of Corralitos in March, about the apple and pear blossoms coming down in showers outside her doorstep, and about her fighting for her last breaths on an old walnut bed just a few steps away from all that loveliness.

What to make of her story? I'm not sure the language even has words that are adequate to the task. But I think what Sarah's story tells us is that there were in fact heroes in the Donner Party, and that heroes are sometimes the most ordinary-seeming people. It reminds us that as ordinary as we might be, we can, if we choose, take the harder road, walk forth bravely under the indifferent stars. We can hazard the ravages of chance. We can choose to endure what seems unendurable, and thereby open up the possibility of prevailing. We can awaken to the world as it is, and, seeing it with eyes wide open, we can nevertheless embrace hope rather than despair. When all is said and done, I think the story tells us that hope is the hero's domain, not the fool's. Because we dare to hope—even when doing so might undo us—we leave the worlds we create behind us, swirling in our wakes, eternal and effervescent with the beauty of our aspirations.

Appendix:

The Donner Party Encampments

November 1846

In early November 1846, the Donner Party scrambled to erect shelters at the south end of Truckee Lake and at Alder Creek five miles to the northeast. As the winter wore on, most of those who belonged to large family groups stayed in their original shelters. As conditions deteriorated, however, many of the single people changed locations, trying to find better opportunities for survival. The following illustrates where each member of the party sheltered during the first frantic weeks. Each person's location and age, where known, is given as of November 12, 1846. Where precise ages are unknown, approximate ages are given in brackets.

THE LAKE CAMP

In the Murphy Cabin

>Levinah Murphy 36
>>John Landrum 16
>>Meriam (Mary) 14
>>Lemuel 13
>>William 10
>>Simon 8
>>Sarah 20 and William Foster 31
>>>Jeremiah (George) 2
>>Harriet Pike 18
>>>Naomi 2
>>>Catherine [1]
>William [28] and Eleanor Eddy [25]
>>James [3]
>>Margaret [1]

In the Breen Cabin

>Patrick [51] and Margaret Breen [40]
>>John 14
>>Edward 14
>>Patrick Jr. 9
>>Simon [8]
>>James 5
>>Peter 3
>>Margaret [1]

Patrick Dolan [35]

Antonio [last name unknown, 23]

In the Keseberg Lean-To

Louis 32 and Philippine Keseberg 23

    Ada 3

    Louis Jr. [5 months]

Augustus Spitzer [30]

Charles Burger [30]

In the Graves-Reed Double Cabin

Jay [23] and Sarah Fosdick 21

Franklin [57] and Elizabeth Graves 46

    Mary Ann 20

    Billy 17

    Eleanor 14

    Lovina 12

    Nancy 8

    Jonathan [7]

    Franklin Jr. [5]

    Elizabeth [1]

Amanda McCutchen [23]

    Harriet [1]

Margret Reed 32

    Virginia 13

    Martha (Patty) 8

James Jr. 5
Thomas 3
Baylis Williams [25]
Eliza Williams [31]
Charles Stanton 35
Luis [Unknown]
Salvador [Unknown]
John Denton [28]
Milt Elliott [28]

In Unknown Shelters

James Smith [25]
Jean Baptiste Trudeau [16]
Noah James [16]

THE ALDER CREEK CAMP

George [62] and Tamzene Donner [45]
Elitha 14
Leanna 11
Frances 6
Georgia 4
Eliza 3

Jacob [56] and Elizabeth Donner [38]
George Jr. 10

Mary 7
Isaac [5]
Samuel [4]
Lewis [3]
Solomon Hook 14
William Hook [12]

Doris Wolfinger [20]
Joseph Reinhardt [30]
Samuel Shoemaker [25]

DIED BEFORE REACHING THE SIERRA NEVADA

Sarah Keyes, May 29, 1846 [70]
Luke Halloran, September 25, 1846 [25]
John Snyder, October 5, 1846 [25]
Mr. Hardcoop, about October 8, 1846 [60]
William Pike, about October 22, 1846 [32]

ARRIVED IN CALIFORNIA BEFORE THE ENTRAPMENT

James Reed 45
Walter Herron [27]
William McCutchen [30]

Acknowledgments

One of the most heartening things about writing a book is how many people always seem to step forward to lend a helping hand. I take it as a positive sign about the state of the world. It suggests that books matter to people, that the bringing forth of books strikes people as a communal responsibility, much as a group of people trapped in an elevator with a pregnant woman about to go into labor might feel they have a common stake in making sure the baby arrives safely.

With that in mind, I would like to express my gratitude to a number of people who have lent a hand in bringing this book wailing into the world. First, I'd like to thank Sarah's great-granddaughter, Kathy Larson, and her husband, Gary, for all the family information and the photographs that they have unstintingly shared with me. Both have been invaluable resources.

I'd also like to pay particular thanks to Kristin Johnson for undertaking the very considerable task of reviewing the manuscript, for her many corrections and constructive comments about it, and for making a number of valuable documents available to me.

A number of other researchers and archivists have also been very helpful. In particular, I am indebted to Juanita D. Larimore and Marilyn Sherwood Kramer for making available Graves family photographs and genealogical information; Judy Malmin for help with the history of Corralitos; Dorothy Folkerts of the Marshall County Historical Society; Judy Russo at Sutter's Fort State Park Archives; the staff of the Bancroft Library in Berkeley; and the staff of the Jean and Charles Schulz Information Center at Sonoma State University.

Jennifer Pooley, my editor at William Morrow, embraced this project enthusiastically from the get-go, and I thank her for that. But I thank her even more for her very perceptive reading of the first draft and for setting me on a more profitable course with the second. And once again I extend my hearty thanks to my agent, Agnes Birnbaum, for tending to the many nitty-gritty details involved in the business side of writing a book.

Above all, I want to thank my wife, Sharon, and my daughters, Emily and Robin, for putting up with my absences while I traveled the country following Sarah; for reading and making many insightful comments on the manuscript; and for always being there, all my pretty chickens.

Chapter Notes

One of the places that I have journeyed while following Sarah is across the landscape of Donner Party literature. I learned early on that it is an uncertain and sometimes treacherous terrain. I have climbed in and out of canyons of conflicting accounts, groped my way through a fog of mythology, and stumbled across arid plains devoid of even a sprig of useful information. But I have found it to be a fascinating and expansive land, well worth traversing.

Even before the last Donner Party survivors arrived at Johnson's Ranch in April 1846, people were beginning to write and to read about the tragedy. The first accounts to appear in two American newspapers just then springing up in California were, for the most part, both overly sensational and inaccurate. They talked of men casually deciding who would live and who would die, of mothers eating the flesh of their babies, of women callously cutting the tongues out of their husbands for a midday meal. These accounts, and others of the same sort in the years and decades that followed, gave birth to an impression of deliberate and widespread moral depravity that has largely defined the Donner Party in the popular imagination to this day.

As I mention in the preface to this book, the first serious attempt to tell the true story was made by Charles F. McGlashan, who corresponded with a large number of Donner Party survivors and then published his *History of the Donner Party,* first in serialized form in the *Truckee Republican* in 1879 and then in book-length form in July of that year. McGlashan's tone and style are sentimental, as is typical of nineteenth-century histories, and many facts have since come to light that undercut parts of his narrative. Nevertheless, it remains peerless simply because McGlashan was able to correspond directly with so many people who lived the tale.

Several other early book-length works also stand as important if sometimes dubious landmarks: J. Quinn Thornton's *Oregon and California in 1848;* Eliza W. Farnham's *California Indoors and Out;* Virginia Reed Murphy's *Across the Plains in the Donner Party;* and

Eliza Donner Houghton's *The Expedition of the Donner Party and Its Tragic Fate*. None of these works is entirely reliable as history, but all are, like McGlashan's *History*, based to some extent on firsthand accounts and are therefore irreplaceable. I have also found Edward Bryant's *What I Saw in California* to be especially valuable. Though Bryant—a cousin of the poet William Cullen Bryant—was not a member of the Donner Party, he traveled just ahead of Sarah and her companions and interacted with a number of them after their arrival in California. As a result, his book is rich in pertinent facts and descriptive detail about the world and the people that Sarah encountered both on the trail and in California.

There are also, of course, many important modern books about the Donner Party. Among those that I have found to be the most valuable are Dale Morgan's superb two-volume anthology of primary sources, *Overland in 1846: Diaries and Letters of the California-Oregon Trail;* Kristin Johnson's anthology of some of the less available accounts of the Donner Party, *Unfortunate Emigrants: Narratives of the Donner Party;* Frank Mullen Jr.'s day-by-day chronicle of the entire saga, entitled, not surprisingly, *The Donner Party Chronicles;* and Donald Hardesty's very interesting *The Archaeology of the Donner Party*.

But all these books, and books in general, represent merely one province in the land of Donner Party literature. Many very valuable firsthand accounts appeared in newspaper articles from around the country shortly after the disaster. A large body of very useful scholarly and semischolarly works has appeared in more than a century's worth of journals that focus on western American history. And most important of all to serious students of the story, a treasure trove of diaries and personal correspondence has been assembled in various libraries and archives in California. Principal among these, the Bancroft Library at the University of California at Berkeley houses the indispensable C. F. McGlashan Papers. The Sutter's Fort archives in Sacramento house another large collection of valuable papers donated by Martha J. (Patty) Reed Lewis. And the Huntington Library in San Marino, California, houses the useful Eliza Poor Donner Houghton Papers.

Early on I was lucky to come across one additional and particularly valuable resource. Kristin Johnson's Web site, "New Light on the Donner Party" at www. utahcrossroads.org/DonnerParty is an extraordinarily rich compendium of facts, dates, narratives, chronologies, links to other sources, digests of current research, biographical data, book reviews, statistics, anecdotes, and news for Donner Party buffs. What makes the site particularly valuable is Johnson's rigorous insistence on accuracy and her emphasis on dispelling myths associated with the tragedy. I cannot recommend it highly enough.

Another excellent resource is Daniel Rosen's Web site at www.donnerpartydiary.com. Rosen offers a detailed, day-by-day chronicle of the major events in the Donner Party saga as well as links to other useful sites.

The Sources section of this book contains a number of additional resources, many of which are referenced in the following notes.

Author's Note

xiii. for no apparent reason: Nancy Graves's frequent bouts of crying at her school in San Jose were noted with concern by a school friend and related to Eliza Donner, who relayed the recollection to McGlashan on August 8, 1879: "I never recall my first schooldays in San Jose without thinking of poor little Nancy G—— who used to cry during school time, and it often seemed to me that her heart would break. . . . We cried with her; and begged her to tell us what troubled her so much; and between sobs and sighs she told us of being at Starved Camp. . . ." The full quotation can be found in Stewart, 313.

xiv. "the slow accretion of national mythology": Stephenson, xvi.

xvi. much to say about her: Georgia Donner remembered Sarah fondly in a letter to McGlashan of October 2, 1879: "[I] have thought so much of Mrs. Fosdick that I have wondered why so little has been said concerning her. . . . She seemed to be very intelligent and sociable" [McGlashan Papers, folder 3].

Prologue

6. emigrants of all ages: For more about the terrible toll that Asiatic cholera took on the emigrants of 1849, see Mattes, 82.

7. "draw you closer and closer": Potter, 202.

8. "you for bread, bread." Ibid. Bryarly was just one of many emigrants who looked for evidence of the Donner Party as they passed through the Truckee area in the years following the tragedy. In June of 1847, when General Stephen Kearny stopped at the site for the purpose of collecting and interring the human remains there, Edwin Bryant was among those traveling with him. Bryant described a macabre scene, quoted in McGlashan, page 328: "I saw two bodies entire, with the exception that their abdomens had been cut open and the entrails extracted. Their flesh had been either wasted by famine or evaporated by exposure to

the dry atmosphere, and they presented the appearance of mummies. Strewn around the cabins were dislocated and broken skulls (in some cases sawed asunder with care, for the purpose of extracting the brains), human skeletons, in short every variety of mutilation."

Chapter One—Home and Heart

13. hung over Steuben Township: The phase of the moon and the time of the sunrise—like all my later mentions of the movements of the sun and the moon— are drawn from data available on the U.S. Naval Observatory's Web site. Online at www.usno.navy.mil.

13. river was black and swollen: Details about the weather and the state of the Illinois River are drawn from the *Illinois Gazette,* April 18, 1846.

15. in their origins and their ways: The histories of the Graves and Fosdick families are drawn from various online genealogical databases.

17. with a puncheon floor: The description of Sarah's family home and many other details of life in Sparland in the 1830s and 1840s are drawn from "Old Settlers of Marshall," from "Mr. Graves and Family," and from Perry Armstrong's oration delivered to the 1879 Old Settlers' Reunion at Lacon, reported by the *Henry Republican,* August 28, 1879, as well as from the remarks of other speakers on the same occasion.

Additional details are from similar occasions recorded in the *Henry Republican* on June 13, 1872, and July 17, 1875. Other details about Sparland, Graves family history, the sale of the Graves property, Levi Fosdick's orchard, the departure for California, and the subsequent journey are derived from Spencer Ellsworth's *Records of the Olden Times: Or Fifty Years on the Prairies.*

20. the bitterly cold winter of 1839–40: The story of Elizabeth Graves's visit to her neighbor is drawn from "Mr. Graves and Family." A few additional facts are drawn from the 1840 and 1850 censuses for Marshall County, Illinois.

25. the genus *Plasmodium*—malaria: More about the pathology of malaria can be found in Hoyt Bleakley's paper "Malaria in the Americas: A Retrospective Analysis of Childhood Exposure" and in Michael Finkel's excellent *National Geographic* article "Bedlam in the Blood."

27. Great Depression of the 1930s: For more about the financial deterioration in the late 1830s, see the chapter entitled "The Financial Panic of 1837" in Bancroft, as well as McLynn, 23.

29. "what a country this might be!": Dana, 187.

30. with plucked beaver fur: My description of Lansford Hastings's appearance depends heavily on an account by emigrant John R. McBride, quoted in Bagley.

31. "abundance of its productions": Hastings, 133.

31. "they are surrounded": Hastings, 114.

31. "with human skulls": Hastings, 116.

32. fifteen hundred dollars in cash: The deed of conveyance for this transaction, dated April 2, 1846, showing the amount is recorded in the Marshall County Courthouse Land Records, Book C, 580–81.

33. and were married: The date of Sarah and Jay's wedding is recorded at the Marshall County Courthouse, certificate #175, Book A, 23.

35. "civilizer that I know of": Hurtado, 71.

35. lots in the new metropolis: For more about Suttersville and Hastings's arrangement with Sutter, see Hook.

36. "by the route just described": Hastings, 137–38.

38. "fifteen or twenty thousand": Lansford Hastings to John Marsh, March 26, 1846, reprinted in Morgan, 39–41.

Chapter Two—Mud and Merchandise

40. to be underwater: Though they later made no reference to it, it is possible that in Iowa the Graves family fell in with a party of some thirty wagons that had started from "Iowa and the country east of it" bound for St. Joe and were delayed by the wet weather and bad roads, as recounted in the *Missouri*

Republican, May 27, 1846, and reprinted in Morgan, 536.

40. "steamers of the largest class": *Illinois Gazette*, April 18, 1846.

42. only about nine months old: The ages of the various members of the Graves family are taken from Kristin Johnson's *Unfortunate Emigrants*, pages 295–96, except for the age of the baby, Elizabeth, which is taken from W. C. (Billy) Graves's "Crossing the Plains in '46" in the *Russian River Flag*, April 26, 1877. Ages here and throughout have been adjusted, where necessary, to account for the appropriate calendar dates, since Johnson uses July 31, 1846, as a baseline.

43. in high spirits: Farnham, reprinted in Kristin Johnson, *Unfortunate Emigrants*, 140.

46. "in which were placed our beds": Virginia Reed, *Across the Plains in the Donner Party*, reprinted in Kristin Johnson, *Unfortunate Emigrants*, 266.

48. early in the twenty-first century: For more about James K. Polk's expansion of executive powers, see Borneman.

48. "saw spooks and villains": DeVoto, 7.

50. "territory which we desire": DeVoto, 191.

50. "War! War!": *Illinois Gazette*, May 16, 1846.

51. better than he did: For more about James Clyman's life, see DeVoto, 54–58.

53. backs of wagons or into tents: For more on the typical sleeping arrangements among the emigrants, see Faragher, 69–70.

55. "alluring them on": *St. Joseph Gazette,* May 8, 1846, roughly when Sarah likely arrived in town.

57. would set her back ten: Information on doctors' rates is from the city of St. Joseph's Web site at www. ci.st-joseph.mo.us/history/medicalrates.cfm.

58. "ten pounds of salt": Hastings, 143.

58. most emigrant families: For more on what kinds of flour and other foodstuffs were available to the emigrants, see Williams, 7–9, and Faragher, 20–24.

61. at the hands of Indians in 1846: Unruh, 185.

63. none other than Lansford Hastings: Bagley, 14.

64. named the river aptly: For much more about the Meek Party, see Boyd, Bassett, and Mariah King's letter on the Oregon History Project's Web site at www. ohs.org/education/oregonhistory.

65. "more swaring then you ever heard": Samuel Parker, quoted on the Oregon History Project's Web site.

65. "and others say hang him": John Herren, quoted in Boyd, 24.

70. scows that served as ferries: See Mattes, 116, for more about the ferries that transported Sarah and emigrants like her across the Missouri.

71. "if it can possibly be avoided": Hastings, 147.

71. "or perhaps forever": Hastings, 144.

Chapter Three—Grass

72. assembling itself in the woods: The party with which the Graves family seems to have set out from St. Joe was called the "Smith party," though just who "Smith" was remains uncertain. The leading candidate is an Oregon-bound emigrant named Fabritus Smith, though he was only about twenty-six in 1846, a somewhat improbable age for the captain of a train full of strong-willed men and women. Some of the families who joined this party—among them the Graveses, the Ritchies, the Starks, and the Tuckers—are listed in the *St. Joseph Gazette*, August 27, 1847, reprinted in Morgan, 731–32.

75. Reason Penelope Tucker and his sons: Many additional tidbits about the Tuckers can be found in Neelands.

77. a great, windswept sea: For more detail about life on the St. Joseph Road and the terrain, see Mattes, 142–49.

80. sat smugly by and watched: The story of Billy Graves's encounter with the resin-weed Indians, including the dialogue, is from his own account in "Crossing the Plains in '46."

82. "exposed to the public gaze": Ellsworth, 112. Many other details of the Black Hawk War, including the abduction of the Hall sisters, are drawn from the same source.

83. "Ye sons of thunder!": Ibid., 121.

86. on either side: During his imprisonment Black Hawk was taken to Washington, D.C., and shown to President Andrew Jackson. In the following months, he was paraded through the major eastern cities, shown off as a kind of war trophy. But the lust for vengeance gradually faded, and Black Hawk became a kind of celebrity in the East, with people lining up in long queues to see him and hear him talk.

87. "look at the bed for her": Virginia Reed's letter to her cousin Mary C. Keyes on July 12, 1846, reprinted in Morgan, 278. Here and elsewhere I have maintained Virginia's spelling, as I think it lends a certain charm and authenticity to her voice.

90. the popular "Virginia Reel": Smith, Mary Ann Harlan. Mary and her father, George, traveled across the plains ahead of the Donner Party that summer.

91. particular notice of each other: It should be noted that Mary Ann later hotly disputed that she and Snyder had been involved in any kind of romance. Writing to Charles McGlashan on July 18, 1879, she took him to task for alluding to the supposed romance in his

History of the Donner Party: "Drop that trash out and insert more useful history. . . . It was all real life of a sterner style" [McGlashan Papers, folder 14]. However, her brother Billy seemed to acknowledge the romance in one of his own letters to McGlashan on March 30, 1879: "As to your 'Romance' I suppose it is as true of the majority of them. But I don't altogether approve [of publishing it]" [McGlashan Papers, folder 16].

92. "wildest and most beautiful scenery": Stanton's letter to Sidney Stanton, June 12, 1846, reprinted in Morgan, 554–57.

93. "to embrace the women": Ibid.

94. "their horses after a hard chase": Tamzene Donner, June 16, 1846, reprinted in Morgan, 561.

94. "the trouble is all in getting started": Ibid.

Chapter Four—Dust

98. had begun to intrigue him: For much more about the life and times of Mariano Vallejo, a thoroughly interesting character, see Rosenus. Many of the details about Casa Grande are drawn from his book, as are the figures on Vallejo's wealth and income.

99. "so many exalted personages": Rosenus, 110.

100. linseed oil and ferric oxide: William Todd listed the materials used to fashion the first bear flag in the *Los Angeles Express*, January 11, 1878.

101. disemboweled and left to die: Bryant recounts the brutal deaths of Cowie and Fowler in chapter 23 of *What I Saw in California.*

102. chasing them off: The killing of Edward Trimble is described in the *Jefferson Inquirer,* July 21, 1846, reprinted in Morgan, 596–99.

105. each night and morning: Mattes, 64.

106. four sets of twins: Lockley, 264.

107. the sexual appetite in particular: Horowitz, 63. For more on sex and contraception in the 1840s, see Horowitz, Brodie, and McCutcheon.

109. on the road to California: We know that Elizabeth Graves brought oil of hemlock with her, because in April of 1879, W. C. (Billy) Graves was present when the site of his family's cabin at Donner Lake was excavated. Among the other items found that day was a sealed tin that still contained traces of the oil. Graves identified it as his mother's.

114. "the misrepresentations of L. W. Hastings": *Oregon Spectator,* June 25, 1846. Reprinted in Morgan, 567.

115. "Hastings published his book of lies": *Oregon Spectator,* June 25, 1846. Reprinted in Morgan, 569.

116. "it may be impossible if you don't": James Clyman's words of advice to Reed were recorded by Ivan Petroff in 1878 and appear in Morgan, 58–59.

116. "so much of a roundabout course": Ibid.

117. five hundred feet above the surrounding desert: Sadly, the spire of Chimney Rock has eroded considerably since 1846 and now stands approximately 350 feet tall.

119. with their rivals the Crow: Charles T. Stanton to Sidney Stanton, July 5, 1846. Reprinted in Morgan, 582–87.

121. killed the day before for the purpose: Ibid.

121. the valley of the North Platte: B. F. E. Kellogg to Preston G. Gesford, July 5, 1846. Reprinted in Morgan, 580–82.

122. from the slough behind the fort: Rosenus, 155.

Chapter Five—Deception

123. the John McCracken family: W. C. (Billy) Graves mentions that only the William Daniels family and the John McCracken family accompanied the Graves family west of Fort Laramie in "Crossing the Plains in '46."

127. "you had better come": B. F. E. Kellogg to Preston G. Gesford, July 5, 1846. Reprinted in Morgan, 580–82.

128. was becoming particularly unpopular: For a further description of Louis Keseberg, see W. C. (Billy) Graves's letter of April 14, 1879, to Charles McGlashan [McGlashan Papers, folder 16].

129. "as far as possible, anticipate them": "Rules for Wives," *Illinois Gazette*, July 25, 1846.

130. "their wives, mere housekeepers": "Rules for Husbands," *Illinois Gazette*, July 25, 1846.

131. "to the day of his death": Lockley, 88.

133. "that led me to this step": Charles Stanton to his brother, May 12, 1846, reprinted in Morgan, 533.

135. along with their husbands and children: The listing of the original Donner Party members and their kinship is derived from the roster in Kristin Johnson's *Unfortunate Emigrants*, 294–98.

137. "selfish adventurer": Tamzene Donner's characterization of Hastings is drawn from a diary entry written by Jessy Quinn Thornton, who traveled in loose association with the Donners up until July 20. The entry is reprinted in Kristin Johnson's *Unfortunate Emigrants*, 22. As Johnson mentions in her own notes, his comment could be the result of hindsight, as at least parts of his diary seem to have been written after the subsequent tragedy.

138. "a faint resemblance to habitable houses": Bryant, 142.

140. "with plenty of water and grass": James Reed to Gersham Keyes, July 31, 1846, reprinted in Morgan, 279–80.

140. "excellent and accommodating gentlemen": Ibid., 279.

144. wood, twisted iron, and gore: Details of the difficult passage through Weber Canyon are drawn

from an account given in *The California Gold Book* by William Wallace Allen and Richard Benjamin Avery and reprinted in Morgan, 418–19.

146. and with their dark destiny: There is some controversy over where and on what date the Graves family overtook and joined the Donner Party. I have chosen to adhere to W. C. Graves's own account that the meeting occurred on August 10. This date was later reinforced by the recollections of both Mary Ann and Lovina Graves. For a detailed discussion of the arguments surrounding this and alternative dates, including August 12 and August 16, see Kristin Johnson's article "When Did the Graves Family Join the Donner Party?" in *Crossroads*, Summer 1996, online at www. utahcrossroads.org/index.html.

146. Reed's Gap: The Miller-Reed Diary, reprinted in Morgan, 262. More of Reed's account of the crossing of the Wasatch can be found in a statement Reed made much later, in two parts, in March and April of 1871, in the *Pacific Rural Press*.

147. not working as hard as they ought: Reed's dissatisfaction with the pace of the roadwork in the Wasatch was reported in the *California Star* on February 13, 1847: "Mr. Reed and others who left the company, and came in for assistance, informed me that they were sixteen days making the road, as the men would not work one quarter of their time."

148. as the grade got steeper: For an account of crossing the steep ridge now known as Donner Hill, see John Breen's statement in Eliza Farnham's *California, Indoors and Out*, reprinted in Kristin Johnson, 142–43.

Chapter Six—Salt, Sage, and Blood

152. and of which gender: Laderman, 22.

152. where they lived, and how well: Ibid., 24.

153. covered with white cloth: Ibid., 31.

154. for transport back home: Ibid., 115.

155. $24 billion a year in the United States: National Casket Retailer Association Web site.

155. $20,000 mahogany caskets: Forest Lawn Memorial Park general price list.

155. photographs of the deceased: "Achieve Immortality with Ink After Life."

157. salt desert that evening: It was Virginia Reed who reported, in *Across the Plains in the Donner Party*, that they set out in the evening.

161. "illusions similar to the mirage": Bryant, 178.

164. "made the mothers tremble": John Breen quoted by Farnham and reprinted in Kristin Johnson's *Unfortunate Emigrants*, 144.

165. "women were mad with anger": The Miller-Reed Diary, reprinted in Morgan, 267.

166. some as few as one a year: Larkin, 166.

166. halitosis and tooth decay: McCutcheon, 162.

167. not mass-produced until the 1880s: "Everyday Mysteries."

169. a nineteenth-century case of road rage: In sorting out the various accounts of what happened between James Reed and John Snyder on October 5, 1846, and trying to arrive at a true version, I have looked for points of consensus among the various reports and discounted any statements that seemed notably slanted, particularly those from members of the Graves or Reed families. That said, however, I do think one observation made by Billy Graves many years later is worthy of note here. Writing to McGlashan on March 30, 1870, he asked rhetorically, "Do you think that a company of over thirty men of a good education and brought up in a civilized country could have been heartless enough to banish and drive out into the wilderness to starve to death a man mearly [*sic*] for accidentally killing a man in self-defense?" [McGlashan Papers, folder 16]. In the end, I am inclined to attribute a bit more culpability to Reed than to Snyder. Reed was widely reputed to have an inclination toward arrogance, and it was he, after all, who pulled a knife, albeit after Snyder had threatened him with a whip stock.

175. Joseph Reinhardt and Augustus Spitzer: That Wolfinger was reputed to be carrying a large amount of money was reported by Houghton, in Chapter 6.

177. significant liberties with the facts: There has been some controversy over whether Reinhardt made a deathbed confession that he had killed Wolfinger as Eddy later reported, because Eddy had presumably already left on the snowshoe expedition when Reinhardt died. However, Leanna Donner, then eleven, was also present at the confession and reported it thirty years later: "Joseph Reinhardt was taken sick in our tent, when death was approaching and he knew there was no escape, then he made a confession in the presence of Mrs. Wolfinger that he shot her husband."

181. "finally concluded to take the California road": George W. Tucker in a letter to McGlashan on April 5, 1879 [McGlashan Papers, folder 51]. Tucker mentions in the same letter that he and his family hoped that the Donner Party had turned around and returned to Truckee Meadows.

182. the other side of the mountains: See Bryant, 52–53, for details of his meeting with Reed on October 28, 1846.

182. bent on owning a piece of it: See Mullen, 188, for mention of Reed's petition to acquire, along with Bryant and Dunleavy, Long Island in the Sacramento River.

183. in place for three or four days: McGlashan, 55.

183. bullet entered Pike's back: That William Pike was shot and died after Stanton rejoined the company,

rather than before as has sometimes been reported, is based on McGlashan's flat-out statement of the fact on page 55.

184. "more than tongue can tell": Mary Murphy to Green T. Lee, May 25, 1847, an excerpt from which appears as a note in Kristin Johnson's *Unfortunate Emigrants*, 43.

186. a month earlier than usual: For evidence that it had snowed as early as October 7 at Donner Pass, see the "Diary of James Mathers," reprinted in Morgan, 243–45. For more about the weather on October 31 and November 1, 1846, see McLaughlin, 3–5.

Chapter Seven—Cold Calculations

195. prairie dogs on a white prairie: McGlashan, 209.

197. "which they did all night": Bryant, 59.

198. difficult for him to use the hand: The exact nature of the wound to George Donner's right hand is described variously in different accounts. See Rarick, 125, for one version.

199. hauled the logs to the site: See Elitha Donner's account, quoted in Hardesty, 59.

203. sixteen people sharing 450 square feet: The physical description of the Murphy cabin in particular is drawn from the archaeological record presented in

Hardesty. Some details about the other two cabins also come from this source.

204. complete the double cabin: Details about the Graves-Reed double cabin are derived from Mary Ann Graves's letter to McGlashan, April 16, 1879 [McGlashan Papers, folder 14]. In a letter to McGlashan, April 15, 1879 [McGlashan Papers, folder 38], Patty Reed later insisted vehemently that Franklin Graves did not build the Reeds' half of the cabin—that the Reeds' teamsters, Eddy, Stanton, and Luis and Salvador did so. However, her assertion is contradicted by a number of others who were there. Patty's sharp words here and elsewhere are reflective of hard feelings between the Reeds and the Graveses that lingered through the lives of all the survivors and well beyond. For a different point of view, it's worth considering what Georgia Donner Babcock had to say in a letter to McGlashan on May 26, 1879: "I'm sorry that [the Reeds] could not come nearer doing right by their fellow travelers than they have done by representing themselves as they have. I am willing that their feelings should be spared as far as possible but expected them to consider the feelings of others" [McGlashan Papers, folder 2].

204. "had minds & wills of their own": Patty Reed to McGlashan, April 15, 1879 [McGlashan Papers, folder 38].

207. ox that had already starved to death: Franklin Graves's sale of the ox to Eddy was reported by Thornton in terms that did not flatter Graves: "He refused to save it for meat, but upon Mr. Eddy's applying to him for it, he would not let him have it for less than $25." It's worth pointing out that Eddy's veracity in many things was questioned by some of his fellow survivors. That said, it is also true that a number of survivors felt that once the harsh realities of the impending disaster began to become clear, the Graveses grew less generous than their neighbors back in Illinois remembered.

210. "very good dog": McGlashan, 123. Reed and McCutchen later engaged in a very pubic argument with Frances H. McDougall, writing on behalf of Mrs. Curtis, about what exactly had happened during and following the dog dinner. This exchange, published in a series of articles in 1871 in the pages of the *Pacific Rural Press*, can be found in Kristin Johnson's *Unfortunate Emigrants*, 181–208.

212. the profile of just such a leader: For more on the qualities that make for good leadership in survival situations, see Leach, 140.

214. and every ounce of them is lethal: "Grizzly Bear Recovery."

217. "on the 31st of last month": Breen, 5. Patrick Breen's diary is probably the single most compelling

document to come out of the Donner Party tragedy. Working in the cold and dark of his cabin at the lake camp, Breen wrote the original on eight sheets of paper folded and trimmed to make thirty-two small pages. I have used a facsimile edition of the version produced by Frederick J. Teggart in 1910.

217. "returned after an unsuccessful attempt": Ibid.

217. "the eve of a snow storm": Ibid.

219. processes then begin to kick in: For a good overview of the physiological effects of hunger, see chapters 2–5 of Russell.

219. greater amounts of salt and other seasonings: For much more about the hunger experiment at the University of Minnesota, see Tucker.

223. "no hopes of finding them alive": Breen, 6.

225. apparent permanent harm to themselves: See Russell, 3–8, for more on "starvation artists" and other survivors of long-term fasts.

225. it looks like this for women: For more about the Harris-Benedict equation and a handy online calculator, "Basal Energy Expenditure: Harris-Benedict Equation," see Cornell University's Web site www-users. med .cornell.edu/~spon/picu/calc/beecalc.htm.

226. simply to maintain her weight: The nutritional value of a Big Mac® can be found on McDonald's Web site at www.mcdonalds.com.

226. 2,679 calories per day: "Dietary Quality and Food Consumption: Dietary Trends from Food and Nutrient Availability Data."

226. 2,158 calories in 1970: Ibid.

Chapter Eight—Desperation

231. "Yours Very Respectfully C.T. Stanton": Charles Stanton to the Donners, December 9, 1846, reprinted in Morgan, 450.

232. "It is our only choice": Mary Murphy quoted in Steed, 15.

233. for Elizabeth Graves to care for: Mary Ann Graves, in a letter to McGlashan dated April 16, 1879, said it was she who convinced Amanda McCutchen to go with the snowshoe party [McGlashan Papers, folder 14].

235. "8 feet deep on the level": Breen, 7.

236. dried beef for each of them: Both Sarah and Mary Ann, in letters written in May 1847, said the snowshoe party carried eight pounds of dried beef each. Sarah's letter first appeared in the *Western Republican* (Lawrenceburg, Indiana) on November 25, 1847.

237. stiff body of Baylis Williams: W. C. (Billy) Graves describes shaving and washing Baylis Williams's body in a letter to McGlashan dated April 1, 1879. However, he incorrectly says Baylis died "about the first of

January" [McGlashan Papers, folder 16]. Patrick Breen had by then already noted Baylis's death on December 17, saying, "Bealis died night before last," which would be the night before the snowshoe party left.

238. precious calories at a furious rate: Mary Ann Graves describes the difficulty the snowshoe party had learning to walk in snowshoes in her letter of April 15, 1879 [McGlashan Papers, folder 14].

241. the sun's unrelenting rays: My information about snow blindness is taken from a variety of online sources, most notably from www.emedicine.com/EMERG/topic759.htm and related pages.

246-47. if we are to remain alive: The information about hypothermia and hyperthermia comes from the Mayo Clinic's Web site at www.mayoclinic.com/health/hypothermia/DS00333.

248. into a kind of death spiral: A discussion of Dr. Hackett's experiments on Denali can be found at the PBS Web site at www.pbs.org/wgbh/nova/denali.

249. "Norwegian fur company among the icebergs": McGlashan, 71–72.

252. who began to fall the farthest behind: My chronology for the snowshoe expedition is based on a variety of sources, including Mary Ann Graves Clarke's account in the *Truckee Republican* of May 17, 1879; additional elements of Mary Ann's account as published

in McGlashan; a statement authored by John Sinclair in 1847 and reprinted in Morgan, 294–95, based on William Eddy's notes and discussions with other survivors; James Reed's account, based again on Eddy and used in J. H. Merryman's "Of a Company of Emigrants in the Mountains of California," in the *Illinois Journal* on December 9, 1847; J. Quinn Thornton's account in *Camp of Death*, 24–39, again based mostly on Eddy; and Patrick Breen's diary, which I have used to correlate weather events observed at the lake camp with those observed by the snowshoe party.

My chronology differs slightly from some others, beginning in particular with December 20. In examining Mary Ann Graves's chronology published in the *Truckee Republican*, I note that she has compressed the events of three days under the heading of a single day, her "Third day." For this one day, in fact, she refers to the passage of the day of December 18, the evening of the eighteenth, the morning of the nineteenth, the evening of the nineteenth, and the morning and day of the twentieth. Therefore, I take her "Fourth day" to be in fact the sixth day, December 21. If so, and if her recollection is otherwise accurate, Stanton was left behind on the Yuba on the morning of December 20. This date is also borne out by Eliza Farnham's account based on interviews with Mary Ann sometime before 1856, and it is also stated explicitly by John Sinclair in his statement of 1847, based on his discussions with survivors not long after the disaster. Mary Ann's chronology grows more confusing as she recounts the subsequent days, when she again compresses multiple days under the headings for single days. Nevertheless, by carefully correlating the survivors' recollections of weather conditions with those recorded contemporaneously by Patrick Breen, it is possible to parse her narrative and arrive at the chronology that I have used. Although it was written some thirty-two years after the events recorded, it is important to note that Mary Ann's is the only detailed, surviving account actually written by a member of the snowshoe party and therefore merits considerable respect despite its obvious deficiencies.

253. hollow stump near the same spot: W. C. (Billy) Graves describes finding Stanton's remains in a hollow stump "about 15 miles along Dutch Flat [Donner Lake Road]" in his March 30, 1879, letter to McGlashan [McGlashan Papers, folder 16]. Assuming that Graves was measuring the distance from the cabins at Donner Lake, this reinforces the notion that Stanton died along the Yuba River before the snowshoe party turned southwest away from the river.

253. his brain would have died: The Mayo Clinic's Web site at www.mayoclinic.com/health/hypothermia/DS00333.

254. "through no organic cause": Leach, 168.

254. life the exception: Parrado, 200.

257. epic blizzards in the high Sierra: For more on the Madden-Julian oscillation, see "Monitoring Intraseasonal Oscillations" at the National Weather Service's Climate Prediction Web site, www.cpc.ncep.noaa.gov/products/intraseasonal.

258. tempers began to flare: For a good summary of the "City of San Francisco" incident, see Rasmussen, B-2.

259. sat motionless until he died: The description of Jacob Donner's death is drawn from McGlashan, 95.

Chapter Nine—Christmas Feasts

261. sisters starve at the lake: Mary Ann Graves's account in the *Truckee Republican* of May 17, 1879.

264. "Your own dear Eleanor": Thornton, 25.

265. were all failing them: For more about the short-term physical and psychological effects of starvation, see Russell, 29–51, and Tucker, 96–127.

265. in Oregon's Coast Range: My information about the James Kim tragedy is drawn from a number of contemporaneous news accounts, all of them online. See Katz, Simon, Yardley, and "Kim Family Search: A Timeline."

269. "What *can* we do?": Mary Ann Graves's account in the *Truckee Republican* of May 17, 1879.

272. beginning to eat at their minds: For more on the psychology of getting lost, see Gonzales, 162–70.

273. had become permanently deranged: Philbrick, 171–75.

275. he did not care: Thornton, 27.

275. they, too, must eat human flesh: Mary Ann Graves's account in the *Truckee Republican* of May 17, 1879.

275. with his daughters at his side: My chronology asserts that Antonio and Franklin Graves died on the day and the night of December 24, respectively. Some published accounts and some primary sources place these events on December 25. Mary Ann Graves, in a letter dated May 22, 1847, states that "Father died on Christmas night at 11 o'clock." I believe that Mary Ann either meant Christmas Eve or was understandably mistaken as to the date. James Reed (based on Eddy's notes) also places these deaths on December 25. Thornton also places them on "Christmas night," but the date he meant by this term seems to be December 24 according to the logic of his narrative. John Sinclair, who talked to Eddy and other survivors after the tragedy, on the other hand, places the deaths, as I do, on December 24. For me, here as in many places, Patrick Breen's meticulous weather observations carry particular weight in reconciling the discrepancies. Most of the

accounts mention in one way or another that Franklin Graves died late at night just as the weather suddenly turned colder and stormier and the rain turned to snow—"in the commencement of the snow storm," as Mary Ann put it herself; "in the beginning of the storm, of cold," as Sarah put it; or as "a most dreadful storm of wind, snow, and hail, began to pour down," as Thornton, based on Eddy, put it.

Patrick Breen's diary entries make it clear that at Donner Lake, just thirty miles to the east and at approximately the same elevation as where Antonio and Franklin Graves died, the precipitation continued to fall as rain through at least midday on December 24. It then turned to snow, either at noon on Christmas Eve or at midnight, depending on whether you take Breen's "about 12 o'clock" to mean A.M. or P.M. Either way, it was on December 24 that colder air arrived at the lake, and by dawn on Christmas Day, according to Breen, it had been snowing "all night and snows yet rapidly." The next day, however, he notes that it ceased snowing on the night of December 25 (Christmas night) and was clear on December 26. So if Franklin Graves died just as colder weather arrived and it began to snow on the ridges above the North Fork of the American River, it does not seem that he could have died as late as midnight on December 25, Christmas night, when in fact the skies were just beginning to clear.

278. experience it before they die: For more on the "hide-and-die" and "terminal burrowing" syndromes associated with the final stages of hypothermia, see Dolinak, 249, and "Hypothermia and Paradoxical Undressing."

279. skipping school on Christmas Day: For this and more about the history of Christmas in New England, see both Frum and Larkin, "Christmas in New England Before 1860."

280. Christmas cards were printed: University of Minnesota Media History Project Timeline.

280. plum puddings, and the singing of carols: See Bodenhamer, 419, for more about the evolution of modern Christmas rituals in the Midwest.

281. "you will want for nothing": "Christmas at Arlington House."

282. "appalling but hope in God Amen": Breen, 8.

282. "you can have all you wish": Virginia Reed Murphy, *Across the Plains in the Donner Party*; and Kristin Johnson's *Unfortunate Emigrants*, 280.

287. never known—fingernail polish: For more about ketosis and other physiological effects of long-term starvation, see Russell, 37–40.

287. "Give me my bone!": Mary Ann Graves's account in the *Truckee Republican* of May 17, 1879. Some accounts have Lemuel Murphy dying after he was offered food, which could only have been human

flesh. Mary Ann's account, however, asserts that only the mouse that Lemuel had eaten alive kept him from dying earlier.

287. above the rim of the canyon: Sarah Foster's recollection of Lemuel's death, in McGlashan, 85, emphasizes clear skies and a bright moon that night, which is corroborated for the night of December 26 by Patrick Breen's diary and by lunar data on the U.S. Naval Observatory's Web site.

Chapter Ten—The Heart on the Mountain

289. a bit less human: See Philbrick, 165–66.

290. rather than resort to cannibalism: Russell, 25.

291. EATING DEAD CHILDREN IS BARBARISM: Ibid., 149.

292. butchering their victims for meat: Ibid. See also "September 8, 1941: Siege of Leningrad Begins."

292. among his apparent victims: Tucker, 8.

292. made soup out of them: Russell, 149–51.

294. hunger-induced psychosis: Tucker, 102.

296. by the time he died: Philbrick, 166.

296. worth of the grisly rations: The estimate that they carried only four days' rations from the "Camp of Death" is Sarah's own, given in her letter of May 23, 1847.

296. the ridge visible on the other side: In tracing the probable route of the snowshoe party, I have followed

the route outlined on Daniel Rosen's very helpful Web site up to the point where they left Sixmile Valley, turned southwest at Emigrant Gap, and made their way down into present-day Onion Valley. After careful study of the survivor accounts, I believe that they did not then immediately continue south and descend into the canyon of the North Fork of the American River, but, trying to regain a southwesterly course, crossed Fulda Creek and moved along the northern side of Blue Canyon, roughly where the Central Pacific Rail Road was later constructed. I think they continued to follow the terrain southwest, following the main canyon of the North Fork but avoiding descending into it until the canyon opened out into the relatively wide expanse of Green Valley, where they descended. I believe they then ascended to the long ridge of Iowa Hill on the southern side of the river. There they could continue traveling southwest, the general direction in which they likely and correctly believed Johnson's Ranch lay. At the far western end of the Iowa Hill ridge, however, they were forced to recross the canyon where the North Fork of the American River turns abruptly to the south, just east of Colfax. Though I am not inclined to follow Thornton in most things, I believe that this route accords with the two crossings that he describes in some detail and that are hinted at in the other accounts. I believe it also accords with the times and distances

outlined by Sinclair and Reed, as well as with the to-
pography described in other accounts, the weather as
tracked in Patrick Breen's journal, and the total amount
of time taken to reach Johnson's Ranch. This route is,
of course, only theoretical, as is any attempt to retrace
the precise route of the snowshoe party, and therefore
in the text I have avoided mentioning specific place-
names when tracing the later wanderings of the snowshoe expedition.

299. "their minds were brooding": Thornton,
28–29.

299. "adversity with unshaken firmness": Ibid., 29.
And Eliza Gregson, who met the snowshoe survivors
when they made it in to Sutter's Fort, corroborates
Thornton concerning the increasing passivity of the
men. See Gregson, 9.

302. friends before family: I am indebted to Kristin
Johnson for pointing out this hierarchy.

304. "you have missed it!": Thornton, 32.

305. "I shall live": Thornton, 33.

306. and tried to die herself: Thornton, 33. W. C.
(Billy) Graves also said that Sarah wanted to die with
her husband, in a letter to McGlashan on March 30,
1879 [McGlashan Papers, folder 14].

306. looking for her and Jay: That Mary was among
those who went back looking for Sarah and Jay Fosdick

is mentioned by W. C. (Billy) Graves in the same letter to McGlashan.

307. "You cannot hurt him now": Gregson, 10.

Chapter Eleven—Madness

311. for seventy miles: "Sir Daniel Gooch."

311. "The Gold Bug": A list of these and other advances that marked the 1840s can be found on the University of Minnesota Media History Project Timeline.

313. dusk gave way to darkness: The various places listed in New York City, descriptions of them, and the information about rail, ferry, and steamship service in and out of the city are all from *Picture of New York in 1846*, 32–160.

313. aid and a hot meal: Ibid., 108.

315. transform California forever: While John Marshall claimed to have found the nugget that set off the Gold Rush, his employee, Peter Wimmer, is by some accounts the one who actually first plucked it out of the mud. Wimmer's wife, Jennie, later famously boiled it in a pot of lye while making soap and thus proved that it was in fact gold when it kept its luster.

317. "Two and three times": Virginia Reed to Mary C. Keyes, May 16, 1847. Reprinted in Morgan, 285.

319. partition day and night: Patty Reed recalled hearing Harriet McCutchen's "terrible screames" in

518 · Chapter Notes

her letter to McGlashan of April 15, 1879 [McGlashan Papers, folder 38].

320. if he pursued the subject: Thornton, 34–35.

320. in order to kill her: W. C. (Billy) Graves, based presumably on Mary Ann and Sarah's own version of events, told a very different version from Eddy's in a letter to McGlashan of March 30, 1879: "Eddy had made remarks which made Mary believe he wanted to kill her to eat" [McGlashan Papers, folder 16].

321. needed to watch their backs: See Philbrick, 172–73, for more about psychic deadening, feral communities, and stress-induced madness.

323. hot stones around a fire: For more on the omnivorous diets of the Maidu and other California Indians, see Kroeber, 409–11, and also Kroeber's "The Food Problem in California," 297–305, in Heizer.

323. the deadliest of poisons: Kroeber, 526.

323. Indians of Northern California: Ibid., 409, 526.

323. though greatly weakened: That Salvador and Luis were alive when the snowshoe party found and subsequently killed them is made clear in both Sarah's letter of May 23, 1847 (see "Survivor Sarah Graves Fosdick"), and Mary Ann's letter of May 22, 1847 (reprinted in Kristin Johnson's Unfortunate Emigrants, 129–31).

325. unnamed Maidu village: The names of various Maidu villages in the area where the snowshoe party

emerged from the high mountains can be found in Kroeber, 394 and plate 37.

326. then licking it off: Ibid., 411.

326. acorn cakes to the strangers: George Tucker to McGlashan, April 5, 1879 [McGlashan Papers, folder 51].

326. fresh grass instead: Thornton, 38.

327. "dreadful to look at": Breen, 10.

328. vomited green bile: Details about the long-term ravages of malnutrition that Sarah and her companions likely suffered are drawn from Russell, 89–90 and 105.

329. as temporary homes: The description of William Johnson's house is based partly on Bryant, 241. See also Steed, 101–4.

331. approaching her family's cabin: Thornton, 39.

331. burst into tears: Ibid.

332. let alone walk: On February 13, 1847, the California Star reported that Sarah and her companions had arrived at Johnson's Ranch, "entirely naked, their feet frostbitten." It seems unlikely that they were completely naked, but they clearly were close to it, and Mary Ann's feet were in fact so swollen with cuts and frostbite that she could not wear shoes for many weeks. Mary Ann Graves talks about the injuries to her feet in her letter of April 16, 1847 [McGlashan Papers, folder 14].

332. their shrunken digestive systems: Thornton, 40.

333. to the hearth of humanity: See Farnham, reprinted in Kristin Johnson's *Unfortunate Emigrants*, 151–52, for a description of the joy with which Mary Ann Graves and others saw light emanating from the Ritchie cabin.

333. "in their heads like stars": Gregson, 9.

Chapter Twelve—Hope and Despair

335. to grind still more wheat: For this and more details of the preparations at Johnson's Ranch, see Daniel Rhoads's statement in Morgan, 325–27.

336. eighteen to twenty hours a day: Trattner, 23.

337. in manufactories with no pay: Ibid., 25.

337. first factory in Rhode Island: Ibid., 26.

338. children under twelve to ten hours: Ibid., 30.

338. under twelve in textile mills: Ibid.

341. "to live or die on them": Breen, 11.

342. the emigrants at Truckee Lake: The date of the First Relief party's departure from Johnson's Ranch is sometimes given as February 5, based on notations in a diary kept by Ritchie and Tucker, the source of many details of my account here. However, two pieces of evidence suggest that this date, and the subsequent dates, are off by one day: First, Patrick Breen records the ar-

rival of the party at the lake camp on February 18, one day earlier than the Ritchie-Tucker diary asserts. And second, various weather events described in the Ritchie-Tucker diary seem to accord much better with weather noted in Breen's diary when a one-day adjustment is taken into account. Therefore I have adjusted the dates of events recorded in the Ritchie-Tucker diary by one day. From February 22 on, the dates of Breen's diary and the Ritchie-Tucker diary agree.

345. "or do you come from heaven?": Daniel Rhoads's statement, reprinted in Morgan, 325–31.

345. "rejoiced to see them rejoice": R. P. Tucker to McGlashan, 1879 [McGlashan Papers, folder 53]. In the same letter, Tucker reports that the first person they saw at the lake was Levinah Murphy.

350. "it maid them all cry": Virginia Reed to her cousin Mary Keyes, May 16, 1847. Reprinted in Morgan, 286.

351. for the return trip: Daniel Rhoads mentions the First Relief eating the last of their beef the first night out in his statement, reprinted ibid., 330.

351. "nigher Pa and somthing to eat": Virginia Reed to Mary Keyes, May 16, 1847, reprinted ibid., 286.

352. "death stared us in the face": Reason P. Tucker's diary entry for February 23, 1847, reprinted ibid., 333.

352. wrapped him in it: The coverlet Tucker abandoned with Denton was worth the considerable sum of twenty dollars. Ibid., 453.

353. would not move on without her: Details of Ada Keseberg's death are from Daniel Rhoads's account, ibid., 325–31; the Ritchie-Tucker diary, ibid., 331–36; and Reason Tucker's 1879 letter to McGlashan [McGlashan Papers, folder 53].

353. "her body to the wolves": Reason P. Tucker's 1879 letter to McGlashan [McGlashan Papers, folder 53].

355. the Californians had surrendered: For more on the "Battle of Santa Clara," see Bryant, 100–101.

355. "the balls that whistled by me": Reed's comments to Sutter about the battle were published in an article by Edwin A. Sherman, entitled "An Unpublished Report of the Battle of Santa Clara," in the *San Francisco Chronicle*, September 4, 1910.

356. "there is no time to delay": McCutchen to Reed, January 27, 1847, Sutter's Fort Archives, item 8-3-308, 48.

358. "they have done so ere this time": Breen, 15.

358. somewhere in the vicinity of Yuba Gap: See King, 95, for more about the location of the meeting of James and Margret Reed. For more details about the meeting itself, see Rarick, 185–86.

358. "gave to all what I dared": Reed's diary of the Second Relief. Reprinted in Morgan, 345.

359. "in raptures": J. H. Merryman's "Narrative of Sufferings of a Company of Emigrants," published in the *Illinois Journal,* December 9, 1847, and reprinted ibid., 298.

360. the Donners' camp at Alder Creek: The dating in Reed's diary appears to be off by one day beginning on March 1. He seems to have incorrectly believed that 1847 was a leap year, for he talks about traveling and arriving at the lake on Monday, February 29, a day that did not exist. Patrick Breen in his diary notes Reed's arrival on Monday, March 1.

362. "There was nothing else": Georgia Donner to McGlashan, May 26, 1879 [McGlashan Papers, folder 2].

363. had left her that morning: The reference to Patty Reed baking bread is from Martha "Patty" Reed Lewis to McGlashan, April 16, 1979 [McGlashan Papers, folder 38]. Cited in Rarick, 195.

363. "snow will be here until June": Breen, 16.

Chapter Thirteen—Heroes and Scoundrels

366. shoes of the emigrants: Some of the other items Reed brought are mentioned in item 8-7-308, 6, at the Sutter's Fort Archives.

367. over the cabins to the east: Thornton, 73.

368. whether she lived or died: See Stewart, 220, and Thornton, 73, for the anecdote of the men joking about playing euchre to see who would get Elizabeth Graves's money.

369. temperature began to plummet: McLaughlin, 103–4.

370. "a great lamentation about the cold": Reed's journal of the Second Relief. Reprinted in Morgan, 347.

370. "for the first fall of the season": James Reed quoting Sutter, in McGlashan, 122.

371. "all covered with water": George Tucker to McGlashan, 1879. Reprinted in Morgan, 324.

371. "has ever been known in Calafornia": Steed, 27.

371. still as deep as twenty feet: Clyman's and Craig's accounts of deep drifts of snow at the crest of the Sierra Nevada are reprinted in McLaughlin, 151–52.

372. "keen as a two-edged knife": Charlotte Brontë, quoted in Barker, 517.

372. perhaps a million lives: Cahill, 221.

372. exceptionally cold weather in the Arctic: See transcripts from the PBS *Nova* broadcast of February 28, 2006, "Arctic Passage." Online at www.pbs.org/wgbh/nova/transcripts/3307_arctic.html.

374. if the air is cold enough: For more on the relationship between precipitation and snowfall accumulations, see McLaughlin, 155–57.

374. forty-one feet of snowfall per winter: See the Sugar Bowl Ski Resort's Web site. Online at www.sugarbowl.com/history.

376. "stared them in the face": Reed's "Narrative of the Sufferings of a Company of Emigrants," *Illinois Journal,* December 9, 1847. Reprinted in Morgan, 299.

377. and the cold grew lethal: See McLaughlin, 104–6, for more about the passage of the cold front through the Sierra Nevada.

377. "nothing ever equaled it": Reed's diary of the Second Relief. Reprinted in Morgan, 349.

389. among survivors of disasters: Foa, 19–20.

389. when subjected to the same trauma: Ibid., 5.

389. than do more mature adults: Van der Kolk, 135.

389. of a parent develop PTSD: Thomas, 38.

390. what you have done: Tangney, 90–110.

390. particularly hostility and aggression: Ibid., 110.

390. norepinephrine levels soar: Thomas, 21–35.

391. hallucinations brought on by the disorder: Kendall Johnson, 48.

391. depression and/or substance abuse: Foa, 7.

391. in the bodies of disaster survivors: Trevisan, 491–94.

392. "how to cry again": Mary Ann Graves, quoted in "Mary Ann Graves, A Heroine of the Donner Party," in *Nevada County Historical Society* 8, July 1954. Reprinted in Kristin Johnson, *Unfortunate Emigrants*, 127.

393. "while they and I live": Morgan, 356. John Stark was perhaps the greatest hero of the Donner Party, almost single-handedly responsible for bringing all of the Breens as well as some of the Graves and Donner children out of the mountains.

396. "save my children!": Thornton, 85.

397. "photographed on my mind": John Breen to McGlashan, April 20, 1879 [McGlashan Papers, folder 11].

398. escape the heartrending scenes: Houghton, chapter 15.

Chapter Fourteen—Shattered Souls

402. a little after noon: Much of what is known about the Fallon expedition comes from an account in the June 5, 1847, edition of the *California Star* reputed to be Fallon's journal. No copy of the journal itself has survived, and there is some controversy as to whether Fallon really authored the alleged journal himself. I have used it judiciously, omitting many of the more lurid details and relying on it primarily for a basic out-

line of events and dates. Much of what it says, however, particularly about Louis Keseberg, is borne out by Keseberg's own statement given to McGlashan in April 1879 and reprinted in McGlashan's book. Other details are substantiated by Reason Tucker's letter to McGlashan, alluded to below.

403. "in this horrible place": Reason P. Tucker to McGlashan, 1879 [McGlashan Papers, folder 53].

407. 2003 in the United States: Arias, 3.

407. historical longevity rates worldwide: The historical statistics here are from Margolis.

407. it was seventy-eight: "Rank Order: Life Expectancy at Birth."

407. their way across the country: Jamison, 1–22.

408. higher degrees of risk taking: Grayson, 153.

408. than that of intact males: Ibid., 154.

408. about as long as men: Laurance.

409. only 14 percent died: All the statistics related to Donner Party mortality can be found in Grayson, 155–58.

409. therefore less heat loss: Ibid., 155.

411. "hury along as fast as you can": Virginia Reed to Mary C. Keyes, May 16, 1847. Reprinted in Morgan, 287.

411. "i want to go to my mother": Mary Murphy, May 25, 1847. Quoted in Steed, 20.

412. "to tell the sufferings of the camp": I am very grateful to Kristin Johnson for providing me with a copy of Sarah Fosdick's letter of May 23, 1847. The complete text of Sarah's letter is now in print in the spring 2008 issue of *Overland Journal*, 24.

414. "and perhaps more than I ought": Mary Ann Graves to Levi Fosdick, May 22, 1847. Reprinted in Kristin Johnson's *Unfortunate Emigrants*, 129–31.

415. "our parents are dead": Hurtado, 206.

416. the hefty sum of $89.93: The ledger listing the debts owed to Sutter by the Graves family and others is at the Sutter's Fort Archives, cataloged as item 8-7-308.

416. negotiations with James Reed by mail: Two letters from Sinclair to Reed at the Sutter's Fort Archives: June 25, 1847, box 8-3-308, 61, and June 28, 1847, box 8-3-308, 62. See also the receipt for forty dollars from James Reed to Sinclair, to be laid out for purchase of young cows to be delivered to Mrs. Fosdick, May 6, 1847, box 8-7-308, 21, and the receipt signed by Sarah Fosdick for forty dollars, being the amount due from James Reed, box 8-7-308, 27.

417. Lansford W. Hastings: Hurtado, 206.

417. mother to her youngest siblings: The list of items Sarah bought at auction is at the Sutter's Fort Archives, box 8-8-308, 122. The spelling book was

almost certainly Noah Webster's *American Spelling Book,* reputedly the second-best-selling book in American history, after the Bible.

Chapter Fifteen—Golden Hills, Black Oaks

418. idly scratching their fleas: For a good sense of how sparsely inhabited the California countryside was in 1846, there is no better read than Bryant's *What I Saw in California.* For a sense of what California was like before the Mexican War, Dana's *Two Years Before the Mast* and also William Henry Thomes's small book, *Recollections of Old Times in California,* are useful.

420. painting it purple and gold: For more about the world Sarah entered when she moved to the Napa Valley, see Wright's "The Early Upper Napa Valley" and also Weber and Neelands.

423. running the mill for Bale: Wright, "The Early Upper Napa Valley," 38–39.

424. On May 10, 1849, he did: See Kristin Johnson's biographical sketch of Mary Ann Graves in *Unfortunate Emigrants,* 126–27, for more about the murder of Edward Pyle. See also Jacob Wright Harlan's account, Harlan, 69–70.

424. in the Napa Valley: "The Early Upper Napa Valley," 38–39. Additional information about the Graves girls' movements around California are from

notes titled "My People," by Elizabeth Cyrus Wright as transcribed by Juanita D. Larimore, to whom I am indebted for providing me with a copy. Other details are from Karl Kortum's unpublished compilation of Graves family history documents. I am grateful to Sarah's great-granddaughter Kathy Larson for making these materials available to me.

426. third son, Alonzo Perry: W. C. (Billy) Graves outlines the sequence of Sarah's first few children in his letter to McGlashan, June 15, 1879 [McGlashan Papers, folder 18].

426. "unworthy of her": Georgia Donner Babcock to McGlashan, May 26, 1879 [McGlashan Papers, folder 2].

427. a dying pig can scream: The anecdote about Sarah, the pig, and the bear is from Wright's "The Early Upper Napa Valley."

428. "like the application of a mustard plaster": London.

429. at the end of the rope: Details of the lynching of William Ritchie come primarily from Gay LeBaron's article "Mob Ruled," from two articles in the *Sonoma Democrat*—"A Disputed Date," December 5, 1891, and an answering piece, "The Disputed Date," December 12, 1891—and from "Sonoma County Sheriff—History and Information."

Chapter Sixteen—Peace

433. wandered down to the bay: Much of my information about life in Corralitos in the 1870s, including information about the location of the old cemetery site, comes from Judy Pybrum Malmin's local history, *Corralitos,* as well as from e-mail exchanges between Judy and myself.

434. Sarah died: Some of the details of Sarah's death are from Karl Kortum's family history compilation.

Chapter Seventeen—In the Years Beyond

435. as best they could: Information about the later lives of the Donner Party members is drawn from a variety of sources, primarily McGlashan and the roster of Donner Party members on Kristin Johnson's "New Light on the Donner Party" Web site. Other important sources include Wright, Weber, King, Steed, Kortum, Neelands, and Fluke.

436. in Santa Rosa in 1907: Kristin Johnson, "New Light on the Donner Party."

436. following their mother's death: Kortum.

436. in 1891 and is buried there: Abercrombie.

436. Sarah's daughters for a time: Ibid.

436. died at the ranch in 1894: Kortum and Kristin Johnson, "New Light on the Donner Party."

436. was buried in Calistoga: For much more about Lovina's marriage and subsequent life, see Wright, "The Early Upper Napa Valley."

437. about it to Charles McGlashan: Nancy Graves to McGlashan, 1879 [McGlashan Papers, folder 55].

437. about thirty feet from a large boulder: The complete story of the recovery of the coins can be found in "Half a Century Buried."

438. Peggy Breen followed him in 1874: For the later lives of the Breens, see McGlashan, 243–44.

439. "have admitted this to be true": James Reed to Gersham Keyes, July 2, 1847. Reprinted in Morgan, 304.

439. died there in 1923: For summaries of the later lives of the Reeds, see McGlashan, 242, and Kristin Johnson, "New Light on the Donner Party."

440. and Leanna in 1930: Dates are from Kristin Johnson, "New Light on the Donner Party."

440. cancer of the jaw in 1878: See McGlashan, 247–252, and Kristin Johnson, "New Light on the Donner Party," for much more about the fates of the Donner children.

440. on Christmas Eve 1859: McGlashan, 243.

441. when he died of a stroke: Kristin Johnson, "New Light on the Donner Party."

441. died in 1934: Ibid.

441. "a drunken sot": Steed, 19–20. Other details, ibid.

441. named Marysville in her honor: McGlashan, 241.

442. died in Tennessee in 1873: Ibid., 242. Other details from Kristin Johnson, "New Light on the Donner Party."

442. "very hungry, eat anything": Stewart, 296.

442. in fact, was hanged: "Astounding Disclosure." *Alta California*, May 10, 1849. Reprinted in "Donner Party Bulletin No. 21," on Kristin Johnson's "New Light on the Donner Party" Web site.

443. a charity hospital in 1895: Kristin Johnson, "New Light on the Donner Party."

443. where she died in 1861: Ibid.

443. at the age of eighty-nine: Ibid.

444. "Hastings was a bad man": Bagley, 21. A number of other facts concerning Hastings's later life are drawn from the same article.

Epilogue

447. "Heart disease was the cause": *Pajaronian*, March 30, 1871.

448. remained in place ever since: Malmin, 31.

459. can still be seen today: Mullen, 49.

461. the peaks ahead of them: McGlashan, 55.

462. had kept oil of hemlock: Ibid., 259–60.

463. paused at the lake camp: See Bryant quoted ibid., 238–39.

463. was formally dedicated: Rarick, 244.

466. won't adapt to *your* needs: This is one of the central theses of Laurence Gonzales's fascinating study of survival psychology, *Deep Survival: Who Lives, Who Dies, and Why.*

Appendix: The Donner Party Encampments

The dates here are drawn from several sources but primarily from the roster at Kristin Johnson's Web site "New Light on the Donner Party," adjusted for the date of November 12, 1846. I would also like to thank Kristin for consulting with and advising me on where each member of the Donner Party was likely living on that date.

Sources

"A Disputed Date." *Sonoma Democrat,* December 5, 1891.

Abercrombie, LeBon G. "Abercrombie Family Tree." Online at www.lgabercrombie.com/p27.htm.

"Achieve Immortality with Ink After Life." Online at http://gizmodo.com/gadgets/death-industry/achieve-immortality-with-ink-afterlife-photo-printed-using-your-ashes-307025.php.

Arias, Elizabeth. "United States Life Tables, 2003." *National Vital Statistics Reports,* vol. 54, no. 14, April 19, 2006. Online at www.cdc.gov/nchs/data/nvsr/nvsr54/nvsr54_14.pdf.

Bagley, Will. "Lansford Warren Hastings: Scoundrel or Visionary." *Overland Journal,* Spring 1994.

Bailey, Eric. "No Proof Donner Clan Were Cannibals." *Los Angeles Times,* January 13, 2006.

Bancroft, Hubert H., ed. *The Great Republic,* vol. 3. Online at www.publicbookshelf.com.

Barker, Juliet. *The Brontës.* New York: St. Martin's, 1996.

Bassett, Karen, Jim Renner, and Joyce White. "Meek Cutoff 1845." Oregon Trails Coordinating Council, 1998. Online at www.endoftheoregontrail.org/oregontrails/meek.html.

Bleakley, Hoyt. "Malaria in America: A Retrospective Analysis of Childhood Exposure." Online at www.princeton.edu/~rpds/downloads/.

Bodenhamer, David J., Robert Graham Barrows, and David Gordon Vanderstel. *The Encyclopedia of Indianapolis.* Bloomington: Indiana University Press, 1994.

Borneman, Walter R. *Polk: The Man Who Transformed the American Presidency.* New York: Random House, 2008.

Boyd, Robert G., and the High Desert Museum. *Wandering Wagons: Meek's Lost Emigrants of 1845.* Bend, OR: High Desert Museum, 1993.

Breen, Patrick. *Diary of Patrick Breen*. Edited by Frederick J. Teggart. Reprint. Silverthorne, CO: VistaBooks, 1996.

Brodie, Janet Farrell. *Contraception and Abortion in Nineteenth-Century America*. Ithaca and London: Cornell University Press, 1997.

Brooks, James F. *Captives and Cousins: Slavery, Kinship, and Community in the Southwest Borderlands*. Chapel Hill: University of North Carolina Press, 2002.

Bryant, Edwin. *What I Saw in California*. Lincoln: University of Nebraska Press, 1985. Facsimile of 1848 edition.

Cahill, Thomas. *How the Irish Saved Civilization*. New York: Knopf, 1996.

Chaffin, Tom. *Pathfinder: John Charles Frémont and the Course of American Empire*. New York: Hill and Wang, 2002.

"Christmas at Arlington House." National Park Service. Online at www.nps.gov/arho/historyculture/christmas.htm.

Cohen, Betsy. "Research Unearths No Donner Cannibals." Missoulian.com. Archived online at

www.missoulian.com/articles/2006/01/13/news/
mtregional/news06.txt.

Dana, Richard Henry, Jr. *Two Years Before the Mast:
A Personal Narrative of Life at Sea.* New York:
Modern Library, 2001.

DeVoto, Bernard. *The Year of Decision, 1846.* Boston:
Houghton Mifflin, 1989.

"Diet Quality and Food Consumption: Dietary
Trends from Food and Nutrient Availability Data."
USDA Economic Research Service Web site.
Online at www.ers.usda.gov/Briefing/DietQuality/
Availability.htm.

Dolinak, David, Evan W. Matshes, and Emma O. Lew.
Forensic Pathology: Principle and Practice. New
York: Elsevier Academic Press, 2005.

"Donner Party Legend Remains Unproven." University of Oregon Public and Media Relations press
release. January 12, 2006. Online at http://waddle
.uoregon.edu/?id=433.

Ellsworth, Spencer. *Records of the Olden Time: Or
Fifty Years on the Prairies.* Lacon, IL: Home Jour-
nal Steam Printing Establishment. Reprint, Henry,
IL: M&D Printing, 1992.

"Everyday Mysteries." U.S. Library of Congress Web site. Online at www.loc.gov/rr/scitech/mysteries/tooth.html.

Faragher, John Mack. *Women and Men on the Overland Trail.* New Haven: Yale University Press, 1979.

Farnham, Eliza W. "Narrative of the Emigration of the Donner Party to California in 1846." In *California Indoors and Out.* New York: Dix, Edwards, 1856.

Field, James. "The Diary of James Field." Series in *Willamette Farmer* (Portland, OR), April–August 1879.

Finkel, Michael. "Bedlam in the Blood." *National Geographic*, July 2007.

Fluke, Lloyd, ed. *Calistoga Centennial 1959: Honoring the Founding of Calistoga by Sam Brannan in 1859.* Napa, CA: Don Crawford Associates, 1959.

Foa, Edna B., Terence M. Keane, and Matthew J. Friedman, eds. *Effective Treatments for PTSD.* New York: Guilford Press, 2000.

Forest Lawn Memorial Park general price list. Online at http://forestlawn.net/pricing/index.asp.

"Franklin Ward Graves." *Marshall County Republican,* November 21, 1867.

Frémont, John Charles. *Narratives of Exploration and Adventure.* Edited by Allan Nevins. New York: Longmans, Green, 1956.

Frum, David. "'The Lord of Misrule' Is Coming to Town." *National Post,* December 23, 2006. Online at www. davidfrum.com/archive.asp?YEAR=2006&ID=392.

Gonzales, Laurence. *Deep Survival: Who Lives, Who Dies, and Why.* New York: Norton, 2005.

Graves, W. C. "Crossing the Plains in '46." *Russian River Flag.* April 26, 1877.

Grayson, Donald K. "Differential Mortality and the Donner Party Disaster." *Evolutionary Anthropology* 2, 1994.

Gregson, Eliza. "The Gregson Memoirs: Mrs. Eliza Gregson's 'Memory.'" *California Historical Society Quarterly,* June 1940. Also online at http://memory .loc.gov/ammem/index.html.

"Grizzly Bear Recovery." U.S. Fish and Wildlife Service Web site. Online at www.fws.gov/ mountain%2Dprairie/species/mammals/grizzly.

Guralnik, J. M., J. L. Balfour, and S. Volpato. "The Ratio of Older Men to Women: Historical Perspectives and Cross-National Perspectives." *Aging*, April 2000, 65–76.

"Half a Century Buried, the Lost Treasure of the Donner Lake Party Found at Last, Relics of a Historic Tragedy." *San Francisco Examiner*, May 24, 1891.

Hardesty, Donald L. *The Archaeology of the Donner Party.* Reno: University of Nevada Press, 1997.

Harlan, Jacob Wright. *California '46 to '88.* San Francisco: Bancroft, 1888. Online at the American Memory, http://memory.loc.gov/ammem/index.html.

Hastings, Lansford W. *The Emigrants' Guide to Oregon and California.* Reprint, Bedford, MA: Applewood Books, 1994.

Hawkins, Bruce R., and David B. Madsen. *Excavation of the Donner-Reed Wagons: Historic Archaeology Along the Hastings Cutoff.* Salt Lake City: University of Utah Press, 1990.

Heizer, R. F., and M. A. Whipple. *The California Indians: A Source Book,* 2d ed. Berkeley: University of California Press, 1971.

"History of the Donner Party: A Tragedy of the Sierras, No. 24. Mary Graves' Version." *Truckee Republican,* May 17, 1879.

Holmes, Kenneth L. *Covered Wagon Women: Diaries and Letters from the Western Trails, 1850.* Lincoln: Bison Books/University of Nebraska Press, 1996.

Hook, Eileen. "Suttersville—a Pipe Dream at Best." *Dogtown Quarterly,* Fall 1994.

Horowitz, Helen Lefkowitz. *Rereading Sex: Battles Over Sexual Knowledge and Suppression in Nineteenth-Century America.* New York: Knopf, 2002.

Houghton, Eliza P. Donner. *The Expedition of the Donner Party and Its Tragic Fate.* Online at www.gutenberg.org/etext/11146.

Howard, Thomas Frederick. *Sierra Crossings: First Roads to California.* Berkeley: University of California Press, 2000.

Hurtado, Albert L. *John Sutter: A Life on the American Frontier.* Norman: University of Oklahoma Press, 2006.

"Hypothermia and Paradoxical Undressing." *Wilderness Medicine Newsletter.* Online at http://

wildernessmedicinenewsletter.wordpress.com/2007/
02/07/hypothermia-paradoxical-undressing/.

Jamison, Dean T., Joel G. Breman, Anthony R. Measham, George Alleyne, Mariam Claeson, David B. Evans, Prabhat Jha, Anne Mills, Philip Musgrove, eds. "Accomplishments, Challenges, and Priorities." *Priorities in Health*, 1–22. New York: Oxford University Press, 2006.

Jeffrey, Julie Roy. *Frontier Women: Civilizing the West?* Revised edition. New York: Hill and Wang, 1979.

Johnson, Dorothy. *Some Went West.* Lincoln: Bison Books/University of Nebraska Press, 1997.

Johnson, Kendall. *Trauma in the Lives of Children.* Alameda, CA: Hunter House, 1989.

Johnson, Kristin. "New Light on the Donner Party." Online at www.utahcrossroads.org/DonnerParty/.

———. "Survivor Sarah Graves Fosdick." *Overland Journal*, Spring 2008.

———. *Unfortunate Emigrants: Narratives of the Donner Party.* Logan: Utah State University Press, 1996.

Kamler, Kenneth. *Surviving the Extremes: What Happens to the Body and Mind at the Limits of Human Endurance.* New York: Penguin, 2005.

Katz, Leslie. "James Kim Found Deceased." CNET News.com, December 7, 2006. Online at www.news.com/James-Kim-found-deceased/2100-1028_3-6141498.html.

"Kim Family Search: A Timeline." KGW.com, December 7, 2006. Online at www.kgw.com/news-local/stories/kgw_120606_news_kim_timeline_.1c6869e.html.

King, Joseph A. *Winter of Entrapment: A New Look at the Donner Party.* Lafayette, CA: K&K Publications, 1994.

King, Mariah Allen. "King Burial and a Letter." April 1, 1846. Transcribed by the Oregon Historical Society. Online at www.ohs.org/education/oregonhistory/.

Kortum, Karl. Unpublished compilation of family history.

Krakauer, Jon. *Under the Banner of Heaven: A Story of Violent Faith.* New York: Anchor, 2004.

Kroeber, A. L. *Handbook of the Indians of California.* New York: Dover, 1976.

Laderman, Gary. *The Sacred Remains: American Attitudes Towards Death, 1799–1883.* New Haven: Yale University Press, 1996.

Larkin, Jack. "Christmas in New England Before 1860." Online at www.osv.org/explore_learn/ document_viewer.php?DocID=2063.

———. *The Reshaping of Everyday Life, 1790–1840.* New York: Harper Perennial, 1989.

Laurance, Jeremy. "Life Expectancy of Women Exceeds Men's for First Time." *Independent* (London), April 7, 2006.

Leach, John. *Survival Psychology.* New York: New York University Press, 1994.

Lebaron, Gaye. "Mob Ruled Sonoma County Is History Lesson." *Press Democrat,* October 15, 1995, A2.

Lockley, Fred. *Conversations with Pioneer Women.* Compiled and edited by Mike Helm. Eugene, OR: Rainy Day Press, 1981.

London, Jack. "Like Argus of the Ancient Times." Online at www.classicreader.com/read.php/bookid .1000/1.

Malmin, Judy Pybrum. *Corralitos.* Self-published, n.d.

Maranjian, Selena. "The Facts of Death." The Motley Fool Web site, October 29, 2004. Online at www.fool .com/investing/small-cap/2004/10/29/the-facts-of-death.aspx.

Margolis, Jonathan. "Our Special Place in History." USA Weekend Web site, December 31, 2000. Online at www.usaweekend.com/00_issues/001231/ 001231newyear.html.

Mattes, Merrill J. *The Great Platte River Road.* Lincoln: Bison Books/University of Nebraska Press, 1969.

McCutcheon, Marc. *Everyday Life in the 1800s.* Cincinnati: Writer's Digest Books, 1993.

McGlashan, C. F. *History of the Donner Party: A Tragedy of the Sierra.* Stanford, CA: Stanford University Press, 1947 (reprint of the revised edition of 1880).

McLagan, Elizabeth. "A Peculiar Paradise: A History of Blacks in Oregon, 1788–1940." Online at http:// gesswhoto.com/paradise-preface.html.

McLaughlin, Mark. *The Donner Party: Weathering the Storm.* Carnelian Bay, CA: MicMac Publishing, 2007.

McLynn, Frank. *Wagons West: The Epic Story of America's Overland Trails.* New York: Grove, 2002.

Meerloo, Jost, and Leo D. Klauber. "Clinical Significance of Starvation and Oral Deprivation." *Psychosomatic Medicine,* vol. 14, no. 6, 1952.

"Mining History and Geology of the Mother Lode." Online at http://virtual.yosemite.cc.ca.us/ghayes/goldrush.htm.

Montgomery, Donna Wojcik. *The Brazen Overlanders of 1845.* Westminster, MD: Heritage Books, 1992.

Morgan, Dale. *Overland in 1846: Diaries and Letters of the California-Oregon Trail.* Lincoln: Bison Books/University of Nebraska Press, 1993.

"Mr. Graves and Family." *Marshall County Republican,* December 5, 1867.

Mullen, Frank, Jr. *The Donner Party Chronicles: A Day-by-Day Account of a Doomed Wagon Train, 1846–47.* Reno: Nevada Humanities Committee, Halcyon Imprint, 1997.

Murphy, Virginia Reed. *Across the Plains in the Donner Party: A Personal Narrative of the Overland Trip to California.* Dillon, CO: Outbooks, 1995.

National Casket Retailers Association. Online at www.
casketstores.com/News.htm.

Neelands, Barbara. *Reason P. Tucker: The Quiet Pioneer.* Napa, CA: Napa County Historical Society,
1989.

O'Brien, Mary Barmeyer. *Heart of the Trail: The Stories of Eight Wagon Train Women.* Guilford, CT:
TwoDot, 1997.

——. *Toward the Setting Sun: Pioneer Girls Traveling the Overland Trails.* Helena, MT: TwoDot,
1999.

Parrado, Nando. *Miracle in the Andes: 72 Days on the
Mountain and My Long Trek Home.* New York:
Three Rivers, 2006.

Peavy, Linda, and Ursula Smith. *Pioneer Women: The
Lives of Women on the Frontier.* Norman: University of Oklahoma Press, 1998.

Philbrick, Nathaniel. *In the Heart of the Sea: The
Tragedy of the Whaleship Essex.* New York:
Penguin, 2001.

*Picture of New York in 1846: With a Short Account
of Places in Its Vicinity; Designed as a Guide to*

Citizens and Strangers; with Numerous Engravings and a Map of the City. New York: C. S. Francis, 1846.

Potter, David M., ed. *Trail to California: The Overland Journal of Vincent Geiger and Wakeman Bryarly.* New Haven: Yale University Press, 1962.

"Rank Order: Life Expectancy at Birth." *The World Factbook.* Online at https://www.cia.gov/library/publications/the-world-factbook/rankorder/2102rank.html.

Rarick, Ethan. *Desperate Passage: The Donner Party's Perilous Journey West.* New York: Oxford University Press, 2008.

Rasmussen, Cecilia. "1952 Blizzard Snared a Train in Its Frigid Grip." *Los Angeles Times,* April 23, 2006. Online at http://articles.latimes.com/2006/apr/23/local/me-then23.

Rosen, Daniel M. "The Donner Party." Online at www.donnerpartydiary.com.

Rosenus, Alan. *General Vallejo and the Advent of the Americans.* Berkeley: Heyday Books/Urion Press, 1999.

Ruggle, Edward. *Picture of New York in 1846.* New York: Homan & Ellis, 1846.

"Rules for Husbands." *Illinois Gazette,* July 25, 1846.

"Rules for Wives." *Illinois Gazette,* July 25, 1846.

Russell, Sharman Apt. *Hunger: An Unnatural History.* New York: Basic, 2005.

Schiraldi, Glenn R. *The Post-Traumatic Stress Disorder Sourcebook.* New York: McGraw-Hill, 2000.

"September 8, 1941: Siege of Leningrad Begins." This Day in History Web site. Online at www.history .com/this-day-in-history.do?id=7014&action=tdih ArticleCategory.

Simon, Scott. "The Heroism of James Kim." December 9, 2006. Online at www.npr.org/templates/story/ story.php?storyId=6602542.

Sioli, Paolo. *Historical Souvenir of El Dorado County, California: With Illustrations and Biographical Sketches of Its Prominent Men & Pioneers.* Oakland, CA: self-published, 1883.

"Sir Daniel Gooch." Online at www.steamindex.com/ people/gooch.htm.

Smith, Mary Ann Harlan. "History of George Harlan." Online at www.harlanfamily.org/georgeh852.htm.

Smith, Ross A. "Oregon Overland: Three Roads of Adversity." Online at http://oregonoverland.com/.

"Sonoma County Sheriff—History and Information." Online at www.sonomasheriff.org/about_history.php.

"South African Development Community Draft Progress Report on Implementation of the SADC Declaration on Gender and Development." May 2007. Online at www.pmg.org.za/docs/2007/071023gender.htm.

Steed, Jack, and Richard Steed. *The Donner Rescue Site: Johnson's Ranch on Bear River.* Sacramento: self-published, 1991.

Stephenson, Michael. *Patriot Battles: How the War of Independence Was Fought.* New York: HarperCollins, 2007.

Stewart, George Rippey. *Ordeal by Hunger: The Story of the Donner Party.* Boston: Houghton Mifflin, 1992.

Tangney, June Price, and Ronda L. Dearing. *Shame and Guilt.* New York: Guilford Press, 2002.

"The Disputed Date—the Time and Place as Given by a Pioneer Resident." *Sonoma Democrat,* December 12, 1891.

"The 1872 Old Setters' Reunion." *Henry Republican,* June 13, 1872.

"The 1875 Old Setters' Meetings." *Henry Republican,* July 17, 1875.

"The 1879 Old Settlers' Reunion." *Henry Republican,* August 28, 1879.

Thomas, Peggy. *Post-Traumatic Stress Disorder.* New York: Lucent Books, 2008.

Thomes, William Henry. *Recollections of Old Times in California: Or, California Life in 1843.* Berkeley: Friends of the Bancroft Library, 1974.

Thornton, J. Quinn. *Camp of Death: The Donner Party Mountain Camp, 1846–47.* Edited by William R. Jones. Golden, CO: Outbooks, 1986.

Trattner, Walter L. *Crusade for the Children: A History of the National Child Labor Committee and Child Labor Reform in America.* Chicago: Quadrangle Books, 1970.

Trevisan, M., E. Celentano, C. Meucci, E. Farinaro, F. Jossa, V. Krogh, D. Giumetti, S. Panico, A. Scottoni, and M. Mancini. "Short-term effect of natural disasters on coronary heart disease risk factors." *Arteriosclerosis* 6, no. 5 (September–October 1986): 491–94.

Tucker, Todd. *The Great Starvation Experiment: The Heroic Men Who Starved So That Millions Could Live.* New York: Free Press, 2006.

"University of Minnesota Media History Project Timeline." Online at www.mediahistory.umn.edu/timeline/1840–1849.html.

Unruh, John D., Jr. *The Plains Across: The Overland Emigrants and the Trans-Mississippi West, 1840–60.* Chicago: University of Illinois Press, 1993.

Van Bruggen, Theodore. *Wildflowers, Grasses and Other Plants of the Northern Plains and Black Hills.* Interior, SD: Badlands Natural History Association, 1992.

van der Kolk, Bessel A., Alexander C. McFarlane, and Lars Weisaeth, eds. *Traumatic Stress: The Effects of Overwhelming Experience on Mind, Body, and Society.* New York: Guilford Press, 2007.

Walker, Dale L. *Bear Flag Rising: The Conquest of California, 1846.* New York: Forge, 1999.

Weber, Lin. *Old Napa Valley: The History to 1900.* St. Helena, CA: Wine Ventures Publishing, 1998.

Williams, Jacqueline. *Wagon Wheel Kitchens: Food on the Oregon Trail.* Lawrence: University Press of Kansas, 1993.

Wright, Elizabeth Cyrus. "My People (Graves Family)." Typescript transcribed from manuscript by Juanita D. Larimore, 2004.

————. "The Early Upper Napa Valley." *Pacific Historian,* Spring 1975, 32–49. Also published as a book by the Sharpsteen Museum, Calistoga, CA, 1991.

Yardley, William. "Man Lost Seeking Help for Family Is Found Dead." *New York Times,* December 7, 2006. Online at www.nytimes.com/2006/12/07/us/07oregon.html?_r=1&oref=slogin.

"Ye Olden Tyme: The 1878 Old Settlers' Reunion." *Henry Republican,* August 22, 1878.

HARPER LUXE

THE NEW LUXURY IN READING

We hope you enjoyed reading
our new, comfortable print size and found it
an experience you would like to repeat.

Well – you're in luck!

HarperLuxe offers the finest in fiction and
nonfiction books in this same larger print size and
paperback format. Light and easy to read, HarperLuxe
paperbacks are for book lovers who want to see
what they are reading without the strain.

For a full listing of titles and
new releases to come, please visit our website:

www.HarperLuxe.com